George Norris

The Early History of Medicine in Philadelphia

George Norris

The Early History of Medicine in Philadelphia

ISBN/EAN: 9783337202477

Printed in Europe, USA, Canada, Australia, Japan

Cover: Foto ©berggeist007 / pixelio.de

More available books at **www.hansebooks.com**

Geo. K. Norris

The

Early History of Medicine

in

Philadelphia.

BY

GEORGE W. NORRIS, M.D.,

AUTHOR OF "CONTRIBUTIONS TO PRACTICAL SURGERY."

LATE SURGEON TO THE PENNSYLVANIA HOSPITAL; PROFESSOR OF CLINICAL SURGERY IN THE
UNIVERSITY OF PENNSYLVANIA; TRUSTEE OF THE UNIVERSITY OF PENNSYLVANIA;
VICE-PRESIDENT OF THE COLLEGE OF PHYSICIANS OF PHILADELPHIA;
PRESIDENT OF THE HISTORICAL SOCIETY OF PENNSYLVANIA;
MEMBER OF THE SOCIÉTÉ MÉDICALE D'OBSERVATION, ETC.

PHILADELPHIA:

1886.

COLLINS PRINTING HOUSE.

Preface.

THE following work on the EARLY HISTORY OF MEDICINE IN
PHILADELPHIA was found among the papers of my Father after his
death in March, 1875. The notes and memoranda which accompany
the manuscript show that it was for the most part written in 1845,
but laid aside for a time, owing to press of active work as a surgeon,
while in later years failure of health prevented the finishing touches
necessary to its completion. Believing that the picture which its
pages present of our early medical men and institutions, as well as of
the customs and manners of our colonists, and the concise but vivid
description of the privations of our Revolutionary Army, and of
the struggles and necessities of its Medical Department, would be
interesting to all antiquarians as well as to the Medical Profession,
I have decided to print it, and have deemed it best to leave it exactly
as it stood, well knowing that my own imperfect acquaintance with
the subject would ill enable me to fill any gaps in it.

WM. F. NORRIS.

(v)

"Quod praecipuum munus Annalium reor, ne virtutes sileantur, utque pravis
dictis factisque, ex posteritate et infamia metus sit."

TACITUS— Annalium, Liber III. Caput LXV.

"Traditional accounts of departed worth, if not combined with some permanent memorial, are commonly short-lived and uncertain."

BACON.

"For nothing else is History
but pickle of antiquity,
where things are kept in memory,
from stincking."

THE WOODSTOCK SCUFFLE.

Inspicere, tanquam in speculum, in vitas omnium
Jubeo ; atque ex aliis sumere exemplum sibi.

TERENTIUS—Adelphi, Act. III. Sc. 3.

(vi)

"Homines enim ad deos nulla re propius accedunt quam salutem homi-
nibus dando."

CICERO—Oratio pro Q. Ligario, Caput XII.

"There PHYSIC fills the space, and far around,
Pile above pile, her learned works abound:
Glorious their aim—to ease the labouring heart;
To war with death, and stop the flying dart;
To trace the source whence the fierce contest grew,
And life's short lease on easier terms renew;
To calm the phrensy of the burning brain;
To heal the tortures of imploring pain;
Or, when more powerful ills all efforts brave,
To ease the victim no device can save,
And smooth the stormy passage to the grave."

CRABBE—The Library.

ERRATUM.

On p. 21, 3d line from bottom, *for* "grandson" *read* "great-grandson."

(viii)

Contents.

BIOGRAPHICAL.

B   (ix)

Contents.

FOUNDATION OF HOSPITALS.

CLINICAL LECTURES.

FOUNDATION OF MEDICAL SCHOOLS.

(xi)

Contents.

I N looking back into the records of the first settlements upon the banks of the Delaware, but few allusions are found either to sickness or physicians. The earliest notice of anything relating to medicine which I have met with is in 1638, in which year it is recorded that **Jan Peterson** from Alfendolft, was employed as "barber" (as surgeons were then denominated), in the settlement of the Swedes, at ten guilders per month.* In the years 1642, 1657, and 1658, Alricks, the Director of the Colony at New Amstel,† notices in his correspondence, the prevalence of "great sickness and mortality," and under the last-mentioned date adds, "Our barber surgeon died, and another well acquainted with his profession is very sick."‡ Afterwards, in 1660, it is represented to the Governor that the company "are much in want of a surgeon," and one

---

* Annals of Pennsylvania, by Samuel Hazard, Phila., 1850, p. 49.
† Now Newcastle.
‡ Annals of Pennsylvania, by Samuel Hazard, Phila., 1850, p. 247.

Peter Typneman offers himself for the post. Besides these passing notices, no mention is made of medicine or its votaries.

Bringing with them few or none who needed a physician's care, these hardy settlers, in case of pressing necessity, sometimes looked for relief to the divine, who, in those days, not unfrequently possessed a smattering of our art, or to the high civil authorities, who also were often dabblers in physic; but as a general rule, nature was their nurse, and temperance their only physician.

Among the English colonists who immediately preceded the arrival of Penn, came John Goodson, "Chirurgeon to the Society of Free Traders." But of him I can find no record, except his removal to Philadelphia after a short residence at Upland. He came from London, was a man of merit, and was probably the first practising physician in Pennsylvania.

With William Penn, himself, in 1682, there arrived three well educated members of the profession, viz., Thomas Lloyd, Thomas Wynne, and Griffith Owen.

Thomas Lloyd, who had been educated at the University of Oxford, and had made considerable progress in the study of literature and science, was a Doctor of Physic. He, however, never practised medicine after his arrival in this country, but took an active part in State affairs, and was made first Deputy Governor of Pennsylvania.

Thomas Wynne, who is said to have been a practitioner in London, also entered into public life, and was elected Speaker of the first Provincial Assembly. He is spoken of by Proud as a "person of note and character." He died in March, 1691.

Griffith Owen was highly esteemed as a preacher among the Friends, and his merit and abilities raised him to several offices of trust; but his practice as a physician, says Proud, "in which he was very knowing and eminent, rendered him of still greater value and importance in the place where he lived." He died in 1717, aged about seventy, universally respected for his professional knowledge as well as for his integrity and public spirit; and though for many years he had the principal practice in Philadelphia, has left no observations concerning the diseases which he met with, or his modes of treatment. The following account of an amputation performed by him is recorded by Thomas Story, and is curious, as well from its being, probably, the first operation of the kind done in the Province, as from its giving some idea of the state of surgery in those days among us. In the firing of a salute, in honor of the landing of William Penn at Chester, in 1699, an accident happened by which a young man had his hand and arm shattered, and the historian adds, "Amputation was resolved upon by Dr. Griffith Owen, the surgeon, and some other skilful persons present. But as the arm was cut off, some spirits in a basin happened to take fire, and being spilt upon the surgeon's apron, set his cloathes on fire; and there being a great crowd of spectators, some of them were in the way, and in danger of being scalded, as the surgeon himself was upon the hands and face; but running into the street, the fire was quenched; and so quick was he, that the patient lost not very much blood, though left in that open bleeding condition."* William Penn, in one of his letters, speaks of him as "tender Griffith Owen, who both sees and feels."

A short time after the arrival of the above-named gentlemen, and in the same year, **Dr. Edward Jones,** a son-in-law of Wynne, emigrated to this Province, and settled in the county of Philadelphia, where he died at an advanced age, leaving a son, a physician, who in after years became the preceptor of Dr. Cadwalader.

---

* Journal of Thomas Story, fol., 1747, p. 240.

The persons above mentioned, constituted at this time the principal medical men of the Province. Gabriel Thomas, in his travels through Pennsylvania, published in 1689, remarks: "Of lawyers and doctors I shall say nothing, because the country is very peaceable and healthy." Indeed, in the history of these early times medicine is scarcely mentioned.

> "We but hear
> Of the survivor's toil in their new lands,
> Their numbers and success."

Pretenders then, as now, abounded, and by them and old Crones, who drew their knowledge from the pages of some one of the many meagre and ill-digested Family Advisers of the day, it seems probable that most of the simple ailments of the colonists were treated. One of our earliest poets, in his story of "Whackum," ridicules in a lively manner those quacks who, in spite of the physicians, retained their influence among the illiterate vulgar. The very slight inducements offered to educated medical men to settle in our wilderness may be judged of from an extract from a letter of Charles Gordon, of the neighboring colony of Jersey, to his brother, Dr. John Gordon, of Montrose. It is of the date of

1685, three years after the arrival of Penn, and in it, he says: "If you desire to come hither yourself, you may come as a Planter, or a Merchant, but as a Doctor of Medicine I cannot advise you; for I hear of no diseases here to cure but some Agues, and cutted legs and fingers, and there is no want of empirics for these already; I confess you could do more than any yet in America, being versed both in Chirurgery and Pharmacie, for here are abundance of curious herbs, shrubs, and trees, and no doubt medicinal ones for making of drugs, but there is little or no employment this way."

In process of time, however, the growing Colony needed medical aid, and in 1711, Dr. John Kearsley, and in 1717, Dr. Thomas Graeme, arrived out from England.

Dr. Thomas Graeme was descended from an ancient family in Perthshire, Scotland, and is spoken of as having been a practitioner in London. He was well educated, of polished address, and literary tastes, and was largely employed as a practitioner for nearly half a century. He possessed the full confidence of his fellow-citizens, but does not appear to have been active in promoting any of our public institutions. As he advanced in life, a deafness, to which he had been in part subject for many years, increased so much as to induce him to decline, in a measure, the practice of physic. The Proprietaries bestowed on him a lucrative office in the Customs, that of Naval Officer for the Port of Philadelphia, wherein he gave great satisfaction, and having a turn for agricultural pursuits, he passed much of his time at his farm in Bucks County, Graeme Park, which his affluent circumstances allowed him the means of greatly embellishing.

Dr. John Kearsley was born in England, emigrated to Pennsylvania in quest of a professional berth, and became one of our most active and valuable citizens. He was much esteemed as a man of eminence and skill in his profession, and for many years was extensively engaged in the practice of both medicine and surgery. He had talents also for public life, and was long one of the representatives for this city in the House of Assembly, and distinguished himself so much in debate where the interests of the Province were concerned, as to have been on several occasions borne from the Assembly to his own house on the shoulders of the people. He possessed considerable skill and taste in architectural matters, and it is said that we are indebted to him for the plan of Christ Church, a building which, at the time of its erection, in point of elegance and taste, surpassed anything of the kind

(15)

in America. He appropriated, by will, a large part of his property to the foundation of Christ Church Hospital, an institution for the support of poor widows of the Episcopal Church, of which communion he was an earnest member. Dr. Kearsley was the preceptor of Zachary, Redman, Bard, and others, who afterwards became distinguished among us. He is stated to have possessed a morose and unhappy temper, and to have treated his pupils with great rigor, requiring of them services of the most menial kind. He died in 1772, in the eighty-eighth year of his age, greatly regretted by our citizens. From a poetical panegyric on him, published in the "Pennsylvania Mercury," for November, 1744, I take the following extract:—

> " Of his great labors and admired skill
> In cure of mortals seized with every ill,
> What safe relief great multitudes have found
> In every Grief, Distemper, Fracture, Wound !
> How far and wide his practice has been spread
> To heal the sick, and almost raise the dead
> To speak at large, the torrent runs too long
> For Plato's numbers and for Tully's tongue.
> Can boys enlarge the sun's refulgent light,
> Or add new lustre, magnitude, or height,

When they, with glasses, in their childish plays,
Make various angles with refracted rays?
So Panegyrics on the Doctor spent
With feeble wings fly short of our intent;
His worthy name with merits compassed round,
Shines bright in Fame, with true-born Honor crowned."

A nephew of the above mentioned gentlemen, **Dr. John Kearsley, Jr.,** was also a physician of note, and in 1769 published an able paper on the Angina Maligna, which had prevailed extensively in the years 1746 and 1760. From the part he took in politics in 1775, he became obnoxious to the Whig party, and, according to Graydon,* had been detected in some hostile machinations. Like his uncle, his temper was unfortunate, and he was impetuous and rash. He was seized at his own door by a party of militia, and in the attempt to resist them received a wound in his hand from a bayonet. Being overpowered, he was placed in a cart and paraded through the streets to the tune of the Rogue's March. The doctor, who lost none of his intrepidity, answered the reproaches and outrages of the mob vehemently, and, by way of retaliation, struck up "God save the King," and continued so infuriate

* Memoirs of a Life passed in Pennsylvania. Harrisburg, 1811.

that he more than once fainted from loss of blood and the violence of his feelings. "I happened," says the writer above
quoted, "to be at the coffee-house when the concourse
arrived there. They made a halt, while the doctor foaming
with rage and indignation, without his hat, his wig dishevelled
and bloody from his wounded hand, stood up in the cart and
called for a bowl of punch. It was quickly handed to him;
when so vehement was his thirst, that he drained it of its con-
tents (to the health of King George) before he took it from his
lips. It had been determined to give him a coating of tar and
feathers; but the tub containing the material, which had been
set in a conspicuous position, was overturned by a friendly
officer. He was finally deposited under guard in the State
House, and a few days afterwards was removed to the jail in
Yorktown, where he became insane, as is said, from political
excitement, and the gross indignities which had been offered
to him. He died at Carlisle, Pa., in November, 1777." The
height to which political animosities were carried at this
period is well shown in the case of this gentleman. He was
not only unanimously expelled from the Society (St. George's),
of which he had long been a member, and the minute directed
to be published in the newspapers, but we are told by Chris-

topher Marshall* that his trial in the church at Carlisle—more than two years afterwards—"disgusted many of the church party in that place, so that they declared against going to that church any more."

Patrick Baird, Chirurgeon, is often mentioned in our records. He appears to have been a man of education, and of some note in the profession, although I can learn but little of him. In 1720, he held an office analogous to that of Port Physician, under the health law of 1700, and had power conferred on him "to board sickly vessels, and examine into the health of the crew and passengers." In 1723, he was chosen Secretary of the Colonial Council, in the place of James Logan, who retired. In 1729, it is mentioned that, along with two other eminent men (Drs. Kearsley and Graeme), he was appointed to make "a personal examination of an individual accused of impotency." In 1740, he was still acting as Secretary to the Council, under Governor Morris. He resigned this office in 1743, on account of failing health, and in appreciation of his services, he received a vote of thanks for his strict probity, diligence, and exactness in the discharge of his

---

* Passages from the Diary of Christopher Marshall, Philada., 1849, p. 163.

duties.[*]   At about this time he left the Province and most probably never returned, for Franklin in his autobiography speaks of his having met him "many years after at his native place, St. Andrews, Scotland."

Nearly contemporary with the first Dr. Kearsley, were Drs. Lloyd Zachary, William Shippen, and Thomas Cadwalader.

Dr. Lloyd Zachary was born in Boston, in the year 1701.   His mother was a daughter of Thomas Lloyd, the first Governor of Pennsylvania, and she dying soon after his birth, he was sent to the care of an aunt in Philadelphia for education.   He entered on the study of physic with Dr. Kearsley, and in 1723 visited Europe to perfect himself in that science, from whence he returned in 1726.   Though of a delicate frame and possessed of fortune, he on his return commenced the practice of his profession, and quickly rose to eminence in it.   His skill and judgment are said to have been great, while the elegance and frankness of his manners engaged the love and esteem of his fellow-citizens.   He was one of the founders of, and first physicians to, the Pennsylvania Hospital; was a founder also of the College of Philadelphia, and by his will, as well as during life, was a liberal benefactor to both.

[*] Colonial Records, vols. iii. and iv.

He never married, and died suddenly September 26th, 1756, while visiting one of his patients. A portrait of him is in possession of the Hospital. He is described by his pupil, Dr. John Jones, "as a person whose whole life had been one continued scene of benevolence and humanity."

Dr. William Shippen was born in Philadelphia, in 1712, and there received his entire education, both literary and medical. Like his friend and colleague, Dr. Zachary, he played an important part in the foundation of several of our literary and charitable institutions, to which also he was a benefactor. He was a Trustee of the College, was one of the first physicians to the Hospital, and Vice-President of the American Philosophical Society. He also contributed largely towards the founding of the College of New Jersey, and bequeathed to it a considerable perpetual annuity. As a practitioner, he had a high reputation, and an extensive business. He died in 1801, having lived to the advanced age of eighty-nine, but relinquished the practice of his profession about the sixtieth year of his life.

Dr. Thomas Cadwalader, a grandson of Dr. Wynne, was born in 1708. After pursuing his medical studies with Dr. Edward Jones, a worthy Welsh physician, settled at

(21)

Merion, on the west side of the Schuylkill, he visited Europe, and having there, under Mr. Cheselden, devoted himself to anatomical pursuits, upon his return to Philadelphia he engaged in dissections and demonstrations for the instruction of his brethren, the first ever made in Pennsylvania; and it is worthy of notice that among those who availed themselves of his teaching was the elder Shippen.

Dr. Cadwalader was one of the physicians appointed to the Pennsylvania Hospital at its commencement, and for more than thirty years continued to be annually re-elected there. He was much celebrated for the polish of his manners and benevolence of disposition; and these, added to his solid attainments in medicine, soon gained for him a distinguished position as a practitioner. His courteous deportment was on one occasion the means of preserving his life. A provincial officer in 1760, laboring under some alienation of mind, left his home one morning armed with a gun, with a determination to kill the first person he should meet. He had not proceeded far before he met Dr. Cadwalader; the doctor bowed politely to the officer, who, though unknown to him, had the appearance of a gentleman, and accosted him with "Good morning, sir; what sport?" The officer answered civilly, engaged in conversation with the doctor, and, as he afterwards declared, was so charmed with his pleasing manner and address, that he had no resolution to carry out his desperate intention. Impelled, however, by the same gloomy disposition that actuated him when he set out, he a few minutes after shot a well known citizen, Mr. Robert Scull, for which crime he was tried and executed.

Dr. Cadwalader was the author of an octavo volume of forty-two pages, entitled "An Essay on the West India Dry Gripes, with the method of preventing and curing that cruel Distemper," to which is added, "An extraordinary Case in Physic." Printed and sold by B. Franklin, 1745. This was one of the first medical publications in America, and the first which has come down to us made in our city. The disease treated of by Dr. C. under the name of Dry Gripes, was in his day a very common affection here, arising from the use of punch. This beverage was made from Jamaica rum, and was the fashionable drink, until pointed out as giving rise to the disease in question, which it did, in consequence of containing poisoning qualities derived from the leaden pipes which were used in its distillation. The common mode of treating the disease was by the employment of crude mercury and drastic purgatives, and the object of the tract in question was to expose the bad effects of these, and to recommend to his fellow practitioners mild cathartics, and the use of opiates, a practice which was afterwards adopted and highly recommended by Dr. Warren, of London. The appendix to this essay is a very interesting case of Mollities Ossium, occurring in a woman aged forty, who in health was five feet high, "but after death, though all her limbs were stretched out straight, was no more than three feet seven inches;" it is stated that the bones in her arms and legs had been so pliable for two years as to be easily bent into a curve, and for several months before her decease they "were as limber as a rag, and would bend any way, with less difficulty than the muscular parts of a healthy person's leg, without the interposition of the bones. An examination of the body was made by Dr. Cadwalader, 1742; and the post-mortem appearance of the different viscera of the chest and abdomen, as well of the bones and joints, are carefully detailed.*

---

* This was one of the first recorded post-mortem examinations made in the American Colonies. The only notice of one made previously to this, of which I am aware, is noted in the early Dutch records of New York, viz., that on the body of the English Governor, Sloughter, in 1691, who was suspected of having died from poison.

Dr. Thomas Bond was born in Maryland, in 1712, and began his medical studies in that State under Dr. Hamilton, a learned practitioner of Calvert County, who had emigrated thence from Scotland in the year 1700. He afterwards travelled in Europe, and spent a considerable time in Paris, in attendance upon the lectures at the Hôtel-Dieu. Upon his return to America he settled in Philadelphia, and commenced business in 1734. Dr. Bond took high rank as a surgeon as well as a physician, and was distinguished for his skill in lithotomy. I find mention made of his having performed that operation successfully in the Pennsylvania Hospital as early as 1756,* and afterwards in 1759, '62, '65, '68, and many succeeding years. In 1765 it is mentioned that he cut three patients, and removed stones weighing one ounce and five drachms (a child aged seven), two and a half drachms, and another "of large size."

---

* This was the first case of stone operated on at the Penna. Hospital (October 29th, 1756), and was soon after the opening of this institution, during its location at Fifth and Market streets. The patient was "a female, from whom was extracted a stone of unusual size." In a report made to the American Medical Association (Transactions, vol. iv. p. 272), Dr. Paul F. Eve mentions that about the year 1760 Dr. Jones, of New York, first performed the operation of lithotomy in America. The first case of Dr. Bond was four years previous to this.

4   (25)

From an old letter which has come under my notice, written by one not belonging to our profession, I extract the following, which is interesting from giving us some idea of Bond's dexterity and worth as an operator, as well as of the state of surgery here, at the early day in which it was written. It is of the date of 1772, and says, "I had the curiosity last week to be present at the hospital, at Dr. Bond's cutting for stone, and was agreeably disappointed, for instead of seeing an operation, said to be perplexed with difficulty and uncertainty, and attended with violence and cruelty, it was performed with such ease, regularity, and success, that it scarcely gave a shock to the most sympathizing bystander, the whole being completed, and a stone of two inches in length, and one in diameter, extracted in less than two minutes." "If," adds the writer, "surgery is productive of such blessed effects, may we not with Cicero justly rank it among the first of arts, and esteem it worthy of the highest culture and encouragement?" Besides enjoying an extensive practice for a period of half a century, this eminent physician was among the foremost in promoting the formation of our useful institutions. He was one of the original members of the American Philosophical Society, and was its first Vice-President. The credit of originating the Pennsylvania Hospital is generally, though erroneously, accorded to Dr. Franklin, for he himself asserts that the suggestion of it is due to Dr. Bond.

When the medical school was originated, the gentleman who proposed and digested the measure, Dr. Morgan, thought it necessary to the design that it should enjoy the aid of Dr. Bond's skill and experience by his delivering a course of clinical lectures in the hospital—the first regular lectures of the kind ever given in America; and his introductory to the course, which is extant, and which will be particularly referred to hereafter, shows how well he was fitted for the task. That they were properly appreciated, is attested by the fact of their being attended in 1766, the first year, by a regular class of thirty students. In 1768 he was appointed Professor of Clinical Medicine in the University of Pennsylvania. Dr. Rush states that it is to Dr. Bond that the city of Philadelphia is indebted for the introduction of mercury into general use in the practice of medicine. He called it emphatically "a revolutionary remedy," and prescribed it in all diseases which resisted the common modes of practice. He gave it freely in the Cynanche Trachealis. Bathing at this period was but little employed in the treatment of acute diseases. Dr. Bond, however, used both the hot and cold baths in the most liberal manner, together with the vapor and warm air baths, both of which he introduced into the practice of our Hospital. Though naturally of a delicate constitution, Dr. Bond, by strict attention to his health, attained the age of seventy-two. He died March 26th, 1784.

Dr. Bond communicated to the "Medical Observations and Inquiries" of London, two papers, the first, giving a minutely detailed history and dissection of an unusual case which fell under his observation in 1753, viz., of a lady who, after an illness of eighteen months, attended with peculiar symptoms, discharged by stool a worm twenty inches long, and an inch in diameter, and soon afterwards died. Upon dissection, the liver was found to be much enlarged and hardened, containing a cavity holding nearly two quarts, filled with bloody water, and a few lumps of coagulated blood. On the side of this cavity was an opening into the hepatic duct, and the biliary ducts were so dilated as readily to admit the end of a common tallow candle. The worm, a figure of which accompanies the case, was annular, "of a red color, and filled with blood in the manner of a leech," and was discharged in two parts. The worm was preserved and sent over to London to William

Hunter, in whose anatomical cabinet Dr. Morgan says that he saw it, ten years afterwards. The second paper (1759) related to the use of Peruvian Bark in scrofula. These, together with an address delivered before the American Philosophical Society in 1782, "On the Rank of Man in the scale of Being, and the conveniences and advantages he derives from the Arts and Sciences," and his introductory lecture already adverted to, are his only literary publications which have come down to us.

His brother, **Dr. Phineas Bond,** who died in 1773, aged fifty-five, was also educated in Maryland, but after studying at Leyden, Edinburgh, Paris, and London, settled in Philadelphia. To this gentleman, along with Thomas Hopkinson, and his brother, is due the credit of originating the scheme of the College, now the University of Pennsylvania. He confined his practice strictly to medical cases, and no medical man in Pennsylvania, says Wistar, "ever left behind him a higher character for professional sagacity, or for the amiable qualities of the heart."

Cadwalader Evans was another distinguished practitioner of those days. He was one of the first pupils of Dr. Bond, and completed his medical studies in England. After his return, he practised physic for a few years in the island of Jamaica, but finally settled in Philadelphia. He was much esteemed as a physician and man of learning, was long attached to the Pennsylvania Hospital, and, according to Franklin, was the originator of the Medical Library in that institution. He bore the reputation of an eminent, candid, and successful physician, whose knowledge was deep and liberal, and improved by an extensive practice, diligent observation, and a penetrating judgment, as well as a zealous promoter of public institutions and literary interests of his State. He died June, 1773, aged fifty-seven, after a lingering illness.

John Redman was born in Philadelphia in 1722, and after receiving his preparatory education at the Academy of the Rev. Mr. Tennent, in Bucks County, commenced his medical studies under Dr. Kearsley. On the expiration of his apprenticeship he went to Bermuda, where he continued for several years to exercise his profession, and thence proceeded to Europe for the purpose of completing his medical education. He spent a year at Edinburgh in attending the lectures of that city, another at Guy's Hospital, London, and the hospitals at Paris, and finally graduated at the University of Leyden, in 1748. At his first setting out in Philadelphia he practised

both surgery and midwifery, and as an accoucheur is stated
to have possessed great skill, but he soon declined these
branches, and confined himself to the practice of physic. In
this he early rose to eminence, but about the fortieth year of
his age he was afflicted with an abscess of the liver, the con-
tents of which were discharged through his lungs, and in
consequence of subsequent delicate health, he withdrew in a
measure from extensive business at a comparatively early
age. Upon the foundation of the College of Physicians in
1786, he was elected president of that body, and for a long
period was one of the physicians of the hospital. From these
institutions, in both of which he ever manifested the deepest
interest, he only retired in consequence of the infirmities of
age. His sole publication, besides an elegant and learned
thesis, "De Abortu," was a defence of inoculation, published
in 1759, in which he urges the use of mercury, in order to prepare the system for the reception of smallpox. Dr. Redman
was a strong advocate for a bold practice, and considered a
more energetic treatment necessary in the cure of diseases in
this climate than in Europe. He bled largely in the yellow
fever of 1762, and gave his influence in support of that practice in 1793. He wrote an account of the yellow fever, as it

(31)

prevailed in Philadelphia in the year 1762, which was communicated to the College of Physicians during the epidemic of 1793. This remained in manuscript till 1865, when, as no description of that epidemic had ever been published, as well as on account of its merits, and as a valuable contribution to our medical history, the tract was judged worthy of publication by that body. He employed mercury freely in all chronic affections, and in diseases of old age he considered small and repeated bleedings as the first of remedies. No physician of his day exerted a more powerful and extensive influence over the practice of medicine in this country than Dr. Redman. "He was faithful and punctual in his attendance upon his patients. In a sick room he possessed the virtues and talents of a specific kind; he suspended pain by his soothing manner, or chased it away by his conversation, which was occasionally facetious, and full of anecdotes, or serious and instructive, according to the nature of the patients' diseases, or the state of their minds. He died in March, 1806, at the age of eighty-six, and until within a few years of his decease, continued to read the latest medical writers, and even warmly embraced some modern doctrines and modes of practice."

A late well-known antiquarian, who had often seen him in

advanced life, gives the following description of him, which is of interest, as showing the dress and appearance of a prominent medical man of the old school. "The doctor, who lived in Second Street near Arch, had retired from practice altogether, and was known to the public eye as an antiquated-looking old gentleman, usually habited in a broadskirted dark coat, with long pocket flaps buttoned across his under dress, wearing, in strict conformity with the cut of his coat, a pair of Baron Steuben's military shaped boots, coming above the knees for riding; his hat flapped before, and cocked up smartly behind, covering a full-bottomed powdered wig, in the front of which might be seen an eagle-pointed nose, separating a pair of eagle black eyes, his lips exhibiting now and then a quick motion, as though at the moment he was endeavoring to extract the essence of a small quid. As thus described in habit and in person, he was to be seen almost daily, in fair weather, mounted on a short, fat, black switch-tailed horse, and riding for his amusement and exercise, in a brisk racking canter about the streets and suburbs of the city."

The same writer gives the following anecdote of him: "He was well known to most of the citizens, and one day, without

5   (33)

ceremony, as he often did, stopped in at a public office, and with a pleasant nod to its occupant, sat himself down to a table to make some memoranda. While thus occupied, he was interrupted by a forward and presuming stranger, who entered, and wanted of him some medical advice gratis. Finding himself thus addressed, he lifted the corner of his wig as usual, and desired the person to repeat his question more loudly, which he did. 'Doctor, what would you advise as the best thing for a pain in the breast?' The wig dropped immediately to its proper place, and after a seemingly profound study for a moment, the doctor replied, 'Oh! ay, I tell you, my good friend, the very best thing I can advise for a pain in the breast is—to consult your physician.'"

Here it may be well to mention that in the last century, in Redman's time, most physicians made their visits on foot, very few but old men riding on horseback, or in little phaetons. In stormy weather they defended themselves with top boots, oiled linen hat covers, and large shoulder capes, hooked around the neck, extending to the knees of the same material, called roquelaires. When umbrellas were introduced, the doctors were the first to patronize them. These articles, although very heavy and clumsy, being made with thick rattan sticks, and covered with oiled silk, were at first ridiculed as effeminate by the newspapers (1772); but some of the professionals, at the head of whom were Drs. Chancellor and Morgan, had the hardihood to carry them, and they soon came into general use.

Contemporary with the latter practitioners whom I have mentioned was **Dr. Adam Thompson,** who settled in Philadelphia in the year 1748. He was originally from Edinburgh, but came here from Prince George's County, Md., whither he had emigrated in the early part of the century. In 1750 he published "A Discourse on the preparation of the body for the Smallpox; and the manner of receiving the infection," a production which was highly spoken of both in Europe and in this country. At the period it was written, inoculation seemed to be on the decline, and Dr. Thompson asserts that the practice of it was so unsuccessful in Philadelphia that many were disposed to abandon it. It was upon the suggestion of the 1392d aphorism of Boerhaave that he was led to prepare his patients by a composition of antimony and mercury, which he states he had constantly employed for twelve years with great success. Dr. Thompson died in the city of New York in September, 1767, and is described as a

(35)

physician of distinguished abilities in his profession, well versed in polite literature, and of unblemished honor and integrity.

William Shippen, Jr., one of the first teachers of Anatomy and Midwifery of which our country can boast, was born in Philadelphia in 1736. He received his primary education at the seminary of Mr. Finley, at Nottingham, and was afterwards placed at the College of New Jersey, where he is said to have been remarked for his knowledge of the Latin language, as well as for his oratorical powers. After graduation he began his medical studies in the office of his father, and at the expiration of three years went to Europe, in order still further to pursue them. His father, writing to a correspondent in England, thus speaks of him: "My son has had his education in the best college in this part of the country, and has been studying physic with me, besides which he has had the opportunity of seeing the practice of every gentleman of note in our city. But for want of that variety of operations, and those frequent dissections which are common in older countries, I must send him to Europe. His scheme is to gain all the knowledge he can in anatomy, physic, and surgery. He will stay in London for the winter, and shall attend Mr. Hunter's anatomical lectures and private dissections, injections, etc., and at the same time go through a course of midwifery with Dr. Smellie; also enter as a pupil in Guy's Hospital. As soon as the season is over, he may go to France, and live with Dr. Leese in Rouen, and there study physic until he can pass an examination and take a degree." While in London he studied Anatomy with John Hunter, and gave much attention to Midwifery, under the direction of his brother William and Dr. McKenzie, a then celebrated teacher in that branch of medicine. He graduated at Edinburgh, the subject of his thesis being, "De Placentæ cum Utero Nexu;" and after an absence of five years he returned home in 1762, and in the autumn of the same year delivered a course of lectures upon Anatomy, with dissections. His announcement of this course was made in a letter to the editor of the leading newspaper of the day, as follows:—

" PHILADELPHIA, November 11th, 1762.

"MR. HALL:—

"Please to inform the public that a course of Anatomical Lectures will be opened this winter in Philadelphia, for the advantage of the young gentlemen now engaged in the study of Physic, in this and the neighbouring provinces, whose circumstances and connections will not admit of their going abroad for improvement to the anatomical schools in Europe; and also for the entertainment of any gentlemen who may have the curiosity to understand the anatomy of the

The Early History of Medicine in Philadelphia.

Human Frame. In these lectures the situation, figure, and structure of all the parts of the Human body will be demonstrated, their respective uses explained, and as far as a course of anatomy will permit, their diseases, with the indications and methods of cure briefly treated of. All the necessary operations in surgery will be performed, a course of bandages exhibited, and the whole conclude with an explanation of some of the curious phenomena that arise from an examination of the gravid uterus, and a few plain general directions in the study and practice of midwifery. The necessity and public utility of such a course in this growing country, and the method to be pursued therein, will be more particularly explained in an Introductory Lecture, to be delivered the 16th instant, at six o'clock in the evening, at the State House, by William Shippen, Jr., M.D.

"The lectures will be given at his Father's house in Fourth Street. Tickets for the course to be had of the Doctor at five pistoles each; and any gentleman who may incline to see the subject prepared for the lectures and learn the art of dissecting, injecting, etc., is to pay five pistoles more."

In February, 1763, the following notice was published:—

"Dr. Shippen having finished on Osteology—the most dry, though the most necessary part of anatomy—will admit gentlemen who want to gratify their curiosity, to any particular lecture. Tickets five shillings."

The number of students who attended these lectures was ten; but he lived to address a class of two hundred and fifty, and to see Edinburgh herself rivalled, if not surpassed, by the school afterwards founded in his native city. The opening of an anatomical room, as might be expected, created much alarm among the good citizens of the place. Mobbing was talked of, and not a little dreaded. Indeed, on several occasions the house in which the dissections were carried on had its windows broken by the populace. In one of these attacks the doctor himself made a narrow escape by passing out through an alley, while his carriage, which stood before the door with its blinds raised, and which was supposed to contain him, received, along with a shower of other missiles, a musket ball through the centre of it. More than once he was obliged to desert his own dwelling and conceal himself, in order to avoid the tyrannical exactions of the people. Several times he addressed the citizens through the public papers, assuring them that the reports of his disturbing private burial-grounds were abso-lutely false, and stating that the subjects he dissected were either of persons who had committed suicide, or such as had been publicly executed; except, he naively adds, "now and then one from the Potter's Field." By his tact and conciliatory deportment, however, joined to the countenance given to him by some respected citizens and the authorities, the excitement against him and his occupations gradually subsided.*

* In December, 1762, the newspapers inform us that the body of a negro who had committed suicide by cutting his throat with a glass bottle, was, after the ver-

(39)

The Early History of Medicine in Philadelphia.

It was not until the time of Shippen that midwifery was generally practised by physicians. Up to his day, except when unusual difficulty occurred, it was left principally in the hands of females. We learn, however, from Dr. Cadwalader, that so early as 1745, a Dr. Spencer was largely engaged in our city, who had returned from Europe "recommended by the famous Dr. Mead, and several other eminent gentlemen of the Faculty of London, as a most judicious and experienced physician and man-midwife;" and three years subsequently, Dr. Thompson also was practising in that branch. In March, 1762, Dr. Shippen delivered, in addition to his anatomical lectures, a special course upon midwifery, the first ever given in the country, the original proposal for which, as possessing much interest, I subjoin.

dict of the coroner's jury, handed over by authority to Dr. Shippen's anatomical theatre. And after that time the bodies of several criminals and suicides were similarly disposed of; and the following paragraph, taken from the Pennsylvania Gazette, shows that even the authorities of the neighboring colony of New Jersey countenanced, and did all in their power to favor, the study of anatomy: "Last Saturday a prisoner was executed at Gloucester, New Jersey, pursuant to his sentence, and his body was sent by order of the Chief Justice to Dr. Shippen's anatomical theatre for dissection."

"DOCTOR SHIPPEN, JUNIOR,

"Proposes to begin his first course on Midwifery as soon as a number of pupils sufficient to defray the necessary expense shall apply. A course will consist of about twenty lectures, in which he will treat of that part of anatomy which is necessary to understand that branch, explain all cases in midwifery—natural, difficult, and preternatural—and give directions how to treat them with safety to the mother and child; describe the diseases incident to women and children in the month, and direct to proper remedies; will take occasion during the course to explain and apply those curious anatomical plates and casts of the gravid uterus at the Hospital, and conclude the whole with necessary cautions against the dangerous and cruel use of instruments.

"In order to make the course more perfect, a convenient lodging is provided for the accommodation of a few poor women, who otherwise might suffer for want of the common necessaries on these occasions, to be under the care of a sober, honest matron, well acquainted with lying in women, employed by the Doctor for that purpose. Each pupil to attend two courses at least, for which he is to pay five guineas. Perpetual pupils to pay ten guineas.

"The female pupils to be taught privately, and assisted at any of their private labors when necessary. The Doctor may be spoke with at his house, in Front Street, every morning between the hours of six and nine; or at his office in Letitia Court every evening."

The above being the first course of lectures upon the subject upon our continent, it may be interesting to notice in detail his Introductory, which I am enabled to do from very full notes of it, which I have had an opportunity of inspecting.

6 (41)

He commences by stating that, having been called to the assistance of a number of women in the country, in difficult labors, "most of which were made so by the unskilful old women about them, and seeing that great suffering to the mothers, accompanied often with loss of life to them, or their offspring, have followed, which could easily have been prevented by proper management, had made him resolve to introduce a course of lectures on that useful and necessary branch of surgery, in order to remedy those terrible evils, and to instruct those women who have had virtue enough to own their ignorance and apply for instruction, as well as those students who are qualifying themselves to practise in different parts of the country with safety and advantage to their fellow-creatures."

Two cases are then related which had fallen under his notice that had been sadly mismanaged. One of these was a natural labor, which was improperly interrupted by the use of instruments, causing the death of the mother; and the other, which occurred near Gloucester, New Jersey, was a shoulder presentation, in which the hand had come down, where two midwives who were in attendance had separated the arm "by knife and scissors," in which he delivered "a full-grown infant by the feet without difficulty in about ten minutes."

"I have reason to believe," says he, "that I shall be able to present each of you with one natural labor at least, and have provided a machine, by which I can demonstrate all kinds of laborious and preternatural labors, and shall give every necessary direction to enable you to manage all cases with the greatest safety to mother and child."

His discourse went on to state the order in which the lectures would be given,* and closed with a brief notice of the rise and progress of the art, from the earliest time down to that of his own teachers, the Hunters and McKenzie. He then points out the qualifications necessary in a man midwife "to make him an adept in his profession, and to gain the good opinion of the female world," recommending to this end a full

---

* The order of these lectures was as follows: 1. On the Bones of the Pelvis. 2. Male and Female Organs. 3. Changes in the Uterus. 4. On the Placenta. 5 and 6. On the Circulation and Nutrition of the Fœtus. 7. On the Signs of Pregnancy. 8. On the Menses. 9. Fluor Albus. 10. On Natural Labours. 11 and succeeding ones. On Laborious or Preternatural Labours, with the Use of Instruments; and concluded by particular lectures On the Diseases of Women and Children within the month, and directions concerning the diet of each, and methods of choosing and making good nurses.

knowledge of its duties, along with "a grave deportment with well-timed conversation, but avoiding religiously any jokes or jests about the patient or profession," and asking pardon of his class for the liberty he is about taking, slyly intimates that the bad habit of drinking may easily be contracted "insensibly by the foolish custom of taking a dram in a cold and wet morning." As to your fees, he adds, "I give you only one admonition, which is, to charge no one extravagantly, and every one in proportion to their abilities, remembering that by giving your services gratuitously to the poor, you will get much from the rich." On the 17th of September, 1765, Dr. Shippen was chosen Professor of Anatomy and Surgery in the College of Philadelphia, and his anatomical lectures, begun as already stated in 1762, were regularly delivered until the fourteenth course, which was in the winter of 1775–6, when they were suspended by the Revolution. In 1770 sundry malicious reports were circulated of his having taken up bodies from the several burying grounds, and the excitement became so great in the community as to make it necessary for him to come out again in the newspapers with a contradiction of it. This piece he closes with the following words: "I have persevered in teaching this difficult and most useful branch of medical knowledge, though attended with very disagreeable circumstances, chiefly from the motive of the public good, and have and always will preserve the utmost decency with regard to the dead; and do again solemnly protest that none of your house or kindred shall ever be disturbed in their silent graves by me or any under my care."

In 1775 he entered the American army, though his lectures were interrupted by his official duties only during the winter of 1776 and '77. In the latter year he was appointed by Congress to succeed Dr. Morgan as Director-General of the Medical Department of the army. Grave charges were preferred against him for his conduct while in this office, of which, however, he was acquitted by Court-Martial in August, 1780, and in October of the same year he was re-appointed to the Directorship. In January, 1781, he resigned the situation, and again taught practical anatomy in a theatre which he had erected at his own expense, previous to his entering the service. For ten years after quitting the army, he continued to practise extensively as surgeon, accoucheur, and physician. Dr. Shippen died at Germantown July 11, 1808, aged seventy-two. As a lecturer, he is described as having been "brilliant," possessing a complete knowledge of his subject, a graceful person, polished manners, with a voice singularly sweet and conciliating. He was remarkable for judgment in selecting what was applicable to elucidating the principles of his subject, plainness in the mode of communicating his thoughts, and a fruitful vein of humor, which he occasionally employed, to awaken the attention of his audience to the subject matter in discussion. His descriptive powers and his fascinating eloquence riveted the attention of his pupils, and impressed with indelible force the lessons he inculcated. Many of his students, after graduation, visited Europe, and all such, says Wistar, "without a single exception, agreed in declaring that they saw no man who was superior to Shippen as a teacher of anatomy, and very few indeed that were equal to him."

John Morgan, deservedly styled the Founder of American Medicine, was born in Philadelphia in the year 1735. His father, Evan Morgan, was a native of Wales, who emigrated early to this Province, where, till his death in 1763, he was a well known and excellent citizen, engaged in mercantile pursuits. Dr. Morgan was his eldest son, and was placed at an early age at Nottingham School, Chester County, under the direction of the Rev. Mr. Finley, a seminary which at that period had a higher reputation than any other in the

John Morgan

Middle Colonies for thorough instruction in the Latin and Greek languages. He afterwards entered the College of Philadelphia, and graduated in the class of 1757, the first that ever received literary honors in that institution. In the last years of his collegiate course, he commenced the study of medicine with Dr. Redman, and in the preface to his "Discourse upon the Institution of Medical Schools in America," he gives an outline of his life from that period to the time of his appointment in the College as Professor of Medicine.

"It is now more than fifteen years since I began the study of medicine in this city, which I have prosecuted ever since, without interruption. During the first six years I served an apprenticeship with Dr. John Redman, who then did, and still continues to enjoy a most justly acquired reputation in this city for superior knowledge and extensive practice in physic. At the same time I had an opportunity of being acquainted with the practice of other eminent physicians in this place; particularly of all the physicians of the hospital, whose prescriptions I put up there above the space of one year. The term of my apprenticeship being expired, I devoted myself for four years to a military life, principally with a view to become more skilful in my profession, being engaged the whole of that time in a very extensive practice in the army, among diseases of every kind. The last five years I have spent in Europe, under the most celebrated masters in every branch of medicine, and have spared no labor or expense to store my mind with an extensive acquaintance in every science that related in any way to the duty of a physician, having in that time expended in this pursuit a sum of money, of which the very interest would prove no contemptible income."

The military life above alluded to was spent with the provincial troops of Pennsylvania in their campaign during the French war. He entered the service as lieutenant and surgeon, but it was in the latter capacity alone that he acted, and served during the whole war. "So great," says Dr. Rush, in his brief notice of him, "was his diligence and humanity in attending the sick and wounded, who were the subjects of his care, that I well remember to have heard it said, 'that if it were possible for any man to merit heaven by his good works, Dr. Morgan would deserve it, for his faithful attendance upon his patients.'"

It was in the year 1760 that he left the army and visited Europe. The friendship of Dr. Franklin, then resident in

London, as agent for the Province of Pennsylvania, introduced him to many prominent men, and gave him the best
opportunities for improving himself under the celebrated medical teachers of that city. While there, he devoted himself
in an especial manner to the lectures and dissections of Dr.
William Hunter.

In November, 1761, he removed to Edinburgh, with the
view of graduating at the school at that place. A copy of a
letter of introduction which he carried from Franklin to Lord
Kames has been preserved, which I am induced here to insert,
as marking the esteem and affection in which he was held
by that eminent philosopher, who had already discovered in
him those germs of promise which time and opportunity
developed.

"May I take the freedom," says Dr. F., "of recommending
the bearer, Mr. Morgan, to your Lordship's protection? He
proposes residing some time in Edinburgh, to improve himself
in the study of physic, and I think will one day make a good
figure in the profession, and be of some credit to the school
he studies in, if great industry and application, joined with
natural genius and sagacity, afford any foundation for the
presage. He is the son of a friend and near neighbour of mine

7 (49)

in Philadelphia, so that I have known him from a child, and am confident the same excellent dispositions, good morals, and prudent behaviour that have procured him the esteem and affection of all that knew him in his own country, will render him not unworthy the regard, advice, and countenance your Lordship may be so good as to afford him."

He likewise carried a letter from Dr. Franklin to Cullen, who received him most kindly. Towards this gentleman, with whom he frequently corresponded and consulted after his return to Philadelphia, he always felt and expressed a most grateful remembrance.

After fulfilling the requirements prescribed at the then famed medical school of Europe, and submitting his thesis, "De Puopoisei," he received the degree of Doctor from the University in that city, in 1763. It was in this thesis that the doctrine was first announced of pus being a true secretion made by the vessels in certain states of inflammation; in it, after showing the difference between his own opinion and that of his predecessors, he thus sets forth his views of the subject. "Hoc mea speciale habet, pus nempe neque in sanguine neque extra vasi generari, sed in ipsis vasis inflammatis; et vasorum mutationes ab inflammatione inductas, esse causas efficientes quae virtute quadam secretoria, pus e sanguine eliciunt."* |

From Edinburgh, Dr. Morgan proceeded to Paris, where he passed a winter in the study of Anatomy, and while there submitted to the Royal Academy of Surgery a memoir on "Suppuration," which was well received, and afterwards exhibited to the same body a number of preparations made by injection and corrosion. This art, which he had learned from the Hunters, was, he informs us, "allowed to be new to them." None of the members present, except M. Morand, who had been in England, and acquainted with Mr. Hunter, having ever seen similar preparations. At their request he

| * That he was the first to announce this doctrine there can be no doubt. The claim to it has been usually awarded to John Hunter; but Mr. Curry, a teacher of anatomy of Guy's Hospital in 1817, after most careful investigation, has adjudged it to Dr. Morgan, who, he says, "discussed the question in his Inaugural Discourse with great ingenuity, and I can find no proof that Hunter taught, or even adopted such an opinion until a considerably later period." (Lond. Med. and Phys. Journ., vol. xxxviii., 1817.) The various views which have prevailed on the origin and formation of pus since that period, form a curious study, and now, after more than a century, Cohnheim (Virchow's Archiv, vol. xxxviii.) has demonstrated that the white corpuscles do actually escape from the intact vessels, and contribute, to a considerable extent, to the formation of pus. |

afterwards presented a memoir "On the Art of making Anatomical Preparations by Corrosion," which procured his admission into that Society. This memoir, as is customary in most learned bodies, was referred to a committee for examination, and M. Sue, the Professor of Anatomy and Chairman of the committee, concludes a very favorable report upon it in these words:—

"On aura donc toujours obligation à Monsieur Morgan d'avoir le premier divulgée les connaissances qu'il a acquises sur cette matière, heureux fruits d'un travail soutenu, avec une constance et une ardeur sans égale. Il ne me reste plus qu'à porter mon jugement sur le mémoire de Monsieur Morgan; c'est pourquoi persuadé qu'il ne peut être que fort utile à l'académie, et augmenter ses richesses anatomiques; je crois qu'elle doit accepter la dédicace que l'auteur lui a faite de son ouvrage, le remercier, et lui donner des marques de sa reconnaissance en lui accordant la place d'Associé à l'Académie; place que sa qualité, son mérite personnel, ses ouvrages, et son dévouement à l'Académie Royale de Chirurgie de Paris, semblent lui avoir acquis à juste titre."*

* In his "Anthropotomie," M. Sue also acknowledges that he received from Morgan the art of making preparations of this sort.

After leaving Paris he travelled through Italy, and while at
Padua, called to pay his respects to the celebrated Morgagni,
to whom he bore letters of introduction, and this venerable
physician, says Rush, was so pleased with the doctor, that
he claimed kindred with him from the resemblance of their
names, and on the blank leaf of the folio edition of his works,
which he presented to him, inscribed with his own hand the
following words :—

"Affini suo, medico praeclarissimo, Johanni Morgan, donat Auctor."

From his MS. Journal, which is now before me, I extract
the following interesting account of this visit. "He received
me," says Morgan, "with the greatest politeness, and showed
me abundant civilities with a very good grace. He is now
eighty-two years of age, yet reads without spectacles, and is
alert as a man of fifty. From his conversation, openness of
behaviour, and winning disposition, I never talked with him
that I did not think I was with my good Dr. Cullen. I found
that he was unacquainted with anatomical preparations made
by corrosion, and showed him a piece of a kidney which I had
injected at Paris, and which was finely corroded, apologizing
at the same time for the state it was in from having been

brought so far. He saw at once the utility of such preparations, and broken as it was, he was highly pleased, and
answered 'ex ungue leonem,' that he saw enough from that
small specimen to convince him of their excellency. Ruysch
when alive, he added, had favoured him with some of his
preparations, but when compared to this, they were 'rudis
indigestaques moles.' "*

While in England he was made a Fellow of the Royal
Society, member of the Belles-Lettres Society of Rome, and
a licentiate of the Royal Colleges of Physicians of London
and of Edinburgh. Thus loaded with literary honors, he
returned to Philadelphia in 1765, to see, as he expresses it in
a letter to Dr. Cullen, "whether, after fourteen years devotion
to medicine, I can get my living without turning apothecary
or practising surgery."

Soon after his arrival, he proposed his plan for connecting
a Medical School with the College of Philadelphia, and intimated his desire to be appointed Professor of the Theory and
Practice of Physic. The project was unanimously approved

* For a perusal of this, as well as for various other original letters, and papers,
I am indebted to the kindness of David Morgan, Esq., of Washington, Pennsylvania.

by the Trustees, who appointed him on the third of May following to the office for which he had applied, a post which he retained until his decease.

Up to the period of Dr. Morgan's settling in our city, it had been usual here, as elsewhere in America, for physicians to practice in all branches of medicine, as well as to prepare and furnish their remedies. This plan he was desirous to change, and recommended a separation of pharmacy and surgery from the practice, expressing his belief that by so doing, the character of the profession would be improved, and each department would be more successfully cultivated. To carry out these views, he determined to confine himself strictly to the practice of physic, refusing all surgical cases, and the furnishing of medicines, and sent his prescriptions to be made up by a gentleman educated both in Pharmacy and Surgery, whom he had brought out from England with him, or if so desired by his patients, they were allowed to choose any other apothecary or surgeon for the operative part.

In urging the utility of separating the duties of the physician and apothecary he remarks: "The paying of a physician for attendance and the apothecary for his remedies apart, is certainly the most eligible mode of practice, both to the patient and practitioner. The apothecary then, who is not obliged to spend his time in visiting patients, can afford to make up medicines at a reasonable price; and it is as desirable as just in itself that patients should allow fees for attendance, whatever it may be thought to deserve. They ought to know what it is they really pay for medicines, and what for physical advice and attendance.

"Nobody, I believe, will deny that the practice of rating medicines at such a price as to include the charge for medicine and attendance is liable to great impositions on the part of ignorant medicasters, too many of whom swarm in every city. Patients who are kept in ignorance of what price medicines are considered separately, and what is the value of physical skill and attendance, naturally think the original cost of medicines, which are comparatively cheap, to be very dear, and undervalue the skill of a physician, his toil of study, and his expense of time and money in his education, which have often amounted to very large sums and to many years spent abroad in quest of knowledge, as if they were of no consideration. The levelling of all kinds of practitioners so much with illiterate pretenders, who have art enough to gain employ, however ill qualified in that of healing diseases, has a tendency to deter persons otherwise of just and liberal sentiments from putting themselves to a further expense to gain knowledge, than what is sufficient to gain money. This is to make a vile trade of physic, instead of a noble profession, which, as it certainly is, so it ought to be esteemed."

The fee fixed upon by Dr. Morgan, in his first setting out in practice, was a pistole for the first visit, as a retaining fee, and a dollar for every visit afterwards. He remarks, however, that it was not his intention to require more than one fee per day, although he might wait on the patient oftener, nor yet every day that he visited once, where a disease of a lingering nature, or requiring particular care, would render his attendance expensive. "A retaining fee I expected to receive from the rich, not from the poor; and had firmly resolved, in no case, to receive more fees than sufficient to pay me for the value of my time and trouble of attendance."

At the date of which we are now speaking, Dr. Morgan had attained nearly the middle period of life. He was a ripe scholar, and possessed an amount of professional learning and experience, probably greater than any medical man who had previously been seen in our Province. His reputation for talent and learning had preceded him, and the most extrava-

8 (57)

gant expectations were formed of his healing powers among the people of his native city. "A venerable gentleman who knew him afterwards very intimately, says the author of an able review of his Discourse, told me that when he was first pointed out to him in the street, he considered it as a high privilege to say 'I have seen him.'" This éclat, added to his graceful and polished manners, procured him at once an introduction into the best practice, and he acquired a large share of business; but the efforts he made to separate the practice from the pharmaceutical parts of the profession gave great offence. Still he persevered in them, and, as he informs us, received fees in cash, agreeably to the custom of the London physicians. His early teachers and contemporaries, the Bonds, Redman, Cadwalader, and Shippen, would not fall into his views, and it was not until some years afterwards that they were generally adopted. This took place on the arrival of Dr. Chovet, a very popular physician, who came here from the West Indies in 1774. Bringing no medicines with him, and being soon engaged in practice, he sent his prescriptions to a druggist. Dr. John Jones, afterwards in 1780, also took up his residence in our city and pursued a like course, and finding it to answer well, he never took a shop. As the doctor came rapidly into esteem, he still further popularized the course which had been begun by Morgan; and the other physicians, by this time seeing the manifest advantages enjoyed by these gentlemen, soon fell into the same practice.

"Had Dr. Morgan succeeded in his design of introducing with the present usages of practice his plan of receiving fees in hand, thereby obviating the necessity of book charges, which, he observes, is never done in any liberal profession, he would have left, says the reviewer previously quoted,* an heirship of the highest value to his successors; valuable, not merely in regard to its pecuniary features, but as abolishing the means, and the incitement to a great many acts, which always, and inevitably tend to depress the profession far below its intrinsic dignity, and relative importance to other professions and callings."

Morgan's zeal was not confined to medicine alone. In other pursuits he was equally persevering and indefatigable, and much of his time was devoted to the improvement of the liberal arts and sciences. He took an active part in the foundation of the American Philosophical Society, and in 1772,

* North American Med. and Surg. Journ., vol. iv. Philadelphia, 1827.

when a project was started to institute a collection in the British West India Islands, in order to raise additional funds for the advancement of general literature in the College, after having received the necessary authority from the Trustees, he undertook a voyage at his own expense for that purpose, and from his exertions an amount equal to two thousand pounds sterling was received.

Upon the breaking out of the Revolutionary War, Dr. Morgan espoused warmly the cause of his country, and in the month of October, 1775, after the removal of Dr. Church, in consequence of traitorous correspondence with the enemy, he was appointed by Congress Director-General to the Military Hospitals, and Physician-in-Chief to the American Army, and immediately joined Washington at Cambridge. Here he found the hospital and army without medicines and appliances, and serious dissensions existing between the officers of his department.

He at once proceeded to reorganize a general hospital, and established rules for the discharge of the various duties of its officers, requiring proofs by examination of the qualifications of the assistants who were to be entrusted with the lives of the sick and wounded soldiery. This requirement gave rise to much dissatisfaction, and five of the surgeon's mates who refused to submit to it were dismissed from the service. By his energy and industry an ample collection of both medicines and stores was made for the General Hospital. His system was economically and prudently arranged; disputes and contentions were repressed; and his personal supervision of, and attention to the sick, was never wanting.

The economy, regularity, and order which he carried into the department committed to his care, together with his opposition to, and reform of abuses that were creeping or had crept into it, raised up to him an enmity among some, at the same time that others, through malignancy or envy, misrepresented his actions, and seized hold of every occasion that times of confusion and public calamity presented, to sully and asperse his character.

To understand the ground of these complaints and misrepresentations, it is necessary to cast a glance at the organization of the department of which he was the head, in the beginning of the revolutionary struggle.

The hospital department was established by act of Congress on the 27th of July, 1775, on a plan reported by a committee of three of that body who had previously been appointed to prepare it.* Upon its organization by Dr. Church, the first Director-General, disputes arose between him and a number of regimental surgeons concerning their duties; the latter claiming the right of having hospitals allowed them, of retaining the sick of their respective battalions under their own care, and of drawing what stores they thought proper from the General Hospital; while the Director, agreeably to the designs of Congress, required them to send the sick to the General Hospital, where they could be well accommodated and provided for, and declined furnishing anything except medicines and instruments to the battalions.

The misunderstandings and disputes thus arising in Dr. Church's time were carried so far as to give rise to courts of inquiry, and regimental hospitals were broken up wherever their surgeons attempted to open them. These difficulties between the two classes of medical officers were aggravated by some of the officers of the army, who, ignorant as they were of the nature and purposes of the General Hospital, justified the regimental surgeons in their attempts to establish themselves in their claims. The abuses practised by these latter in their exorbitant drafts from, and demands upon the

* *Vide* Appendix I.

General Hospital were much talked of previous to Dr. Morgan's appointment, and it had been his first care to look into and correct them.

At the breaking out of the war it must be remembered, along with some of liberal manners and education, and of the best abilities, a number of unlettered and very incompetent medical officers found their way into the army. Many of those first commissioned were never educated to the profession, and were to the last degree ignorant, factious, and turbulent, averse to all subordination and order, and, as is stated by the highest authority, that of Washington, "a disgrace to the profession, the army, and to society."[*]

---

[*] The regimental surgeons and mates, many of whom are designated by Washington, whose judgment and discretion will not be questioned by any, as "very great rascals," were dissatisfied also at having the officers of the General Hospital take equal rank with themselves. There is, adds Washington, a "constant bickering among them, which tends greatly to the injury of the sick, and will always subsist until the regimental surgeons are made to look to the Director-General of the Hospital as a superior." (Letter to the Pres. of Congress, Sept. 24th, 1776. Sparks's Life, vol. iv. p. 116.)

"They are aiming, I am persuaded, to break up the General Hospital and have in numberless instances, drawn for medicinal stores in the most profuse and extravagant manner for private purposes." (Ibid.)

Among some of these dissatisfaction with and enmity to the new Director-General arose, in the first instance, from his requiring proofs of capacity by examination, and by his reform of hospital abuses, and soon acquired additional vigor from his inflexible resolution of being faithful to his trust, and not suffering the stores and instruments collected for the General Hospital to be dissipated among the various regi-

Gordon in his History of the War asserts, that some of the regimental surgeons made a practice of selling recommendations to furloughs and discharges, and states that one of them proved to have done it was drummed out of the army.[*] In another place he adds, "several of the regimental surgeons had no professional abilities, and had never seen an operation of surgery, and were ignorant to a degree scarcely to be imagined." (Vol. ii. p. 115, N. Y., 1801.)

At Cambridge one regimental surgeon drew upon the General Hospital "for above 100 gallons of rum, with wine and sugar in proportion, in the space of six weeks, and from this regiment there was no return of sick made." (Ibid. p. 42.)

"Some of them in time of need deserted the army altogether, and others shunned to attend the regiments to the field of battle." (Ibid. p. 104.)

"At the action of White Plains few of the regimental surgeons were with their respective battalions, in consequence of which many of the wounded bled to death." (Ibid. p. 105.)

---

[*] A surgeon at Harlem was drummed out of his regiment "for selling soldiers certificates that they were unfit for duty." (Diary of Revolution, vol. i. p. 315.)

mental surgeons. This latter course he pursued in obedience
to the resolve of Congress of July 12, 1776, as well as by
repeated directions from the Commander-in-Chief, in order to
oblige the regimental surgeons to send their sick to the General Hospital, and to look for such supplies as they might need
from the proper officers—the Continental druggists.

The necessity for this resolve, to correct the abuses practised in drawing for all sorts of expensive stores, may be
judged of from the fact that the Commissary-General complained that "it required a greater sum of money to answer
their demands than was sufficient to defray the expenses of
all the well men in the army." This measure, though deemed
by Congress, after investigation, absolutely necessary, produced almost universal discontent among the regimental surgeons, who, in consequence of it, poured forth denunciations
both loud and bitter against the chief of the hospital department. On the 9th of October of the same year, a further
resolve of Congress reiterated "that no regimental hospital
should be allowed in the neighbourhood of the General Hospital," a measure which had become imperative from the fact
of the sick being retained in their regiments, without even
reporting them to the Director. This gave rise to additional

9 (65)

ill feeling, and the odium of the measure was again visited upon him.

Having had some years' experience in the English military hospitals during the French war, accustomed as he had been to strict discipline, and aware of its beneficial effects, Dr. Morgan enforced it in carrying out these several orders, at the same time that he corrected minor, though hardly less flagrant abuses. But though able, faithful, and indefatigable in the discharge of his duties, yet all his exertions proved ineffectual. Strict order and economy in the medical as in other departments of the army at that day seem to have been impracticable. The difficulty of supplying the hospitals was as great as that of providing the soldiers with arms, and the stores and assistants that could be furnished by the committee of Congress proved insufficient for the increased demands made upon him. Medicines became daily more and more scarce; bandages, lint, tow, even the most common articles, such as rags, old linen, thread, pins, etc., were not to be procured in sufficient quantity. We find him at this period writing to Rush and Gerry, at that time two of the most prominent members of the medical committee, acquainting them with the wants of the sick, and urging them to direct the attention of Congress to this matter as a serious subject, stating that he must "not be blamed for the impossibility of collecting them," and predicting, too truly, those calamities to the army, if not furnished with them, which afterwards appeared and swept thousands to their graves. Fresh discontent now showed itself, and unjust complaints from this cause, as well as from those before mentioned, again arose, in which enemies, whom he afterwards proved to have been guilty of fraud and peculation, joined. These complaints at last reached the halls of Congress, and were there listened to by some, though no charge was made of his neglect of duty, want of capacity, inattention, or any breach of the resolves of Congress or of general orders.

At about this juncture he was directed by Congress to provide and superintend a hospital at a proper distance from the camp, for the army posted on the east side of Hudson's River, while the director of the flying camp, which had been formed for the protection of New Jersey, was ordered to provide and superintend a similar one for the men on the west side of that river, and to report, not through the Director-General as had been previously ordered by a resolve of July 17th, but to the Congress itself. At this, Dr. Morgan took umbrage, looking upon himself as degraded from the rank of Director-General and Physician-in-Chief to that of Director only, and took the liberty to write to the President of Congress, stating "that he hoped there would be no unnecessary abridgment of rank and authority in his person that was necessary for the head of the department, and was consistent with real usefulness, to which every other consideration should give way." He nevertheless obeyed the order of Congress, looking upon it as a matter for present convenience.

In the northern department of the army, to which Dr. Stringer had been appointed by Congress Chief Physician and Surgeon, independent of Dr. Morgan as Director-General, the sufferings of the sick at this time were great. In a letter to Mr. Samuel Adams, a member of the Medical Committee of Congress at this date, Dr. Morgan describes the state of the army there (derived from a medical officer who had just left it), as "truly deplorable and scarcely credible." "From all I am able to learn," he adds, "everything in the Medical Department in Canada displays one scene of confusion and anarchy; nor has the Congress taken upon itself, or vested any person with a power sufficient to establish a general hospital there. I am not sure that our disgrace and misfortunes in Canada are not owing, in great measure, to the shameful proceedings of the surgeons in spreading the smallpox by inoculation among the soldiery in face of the enemy."*

A clergyman, a witness to their sufferings, forwarded a letter to General Washington at this time, describing the state of the sick at the lakes, in which he says, "Men of considera-tion do not hesitate to speak freely of first-rate characters as the authors of these calamities, some blaming the Congress, some the Surgeon-General of the Army, and some the Direc-tor of the Northern Department." That Dr. Morgan was in no way accountable for the mismanagement of the Northern Department and wants of the sick, there is abundant evidence to prove. He had long before acquainted Congress with their situation, and prayed for relief for them in vain, and although without authority in that department, had sent, at the earnest request of its Director, all he could spare from his own stores, as well as officers to assist them.

Having, by direction of General Washington, put the hos-

---

* This was done without fitting up an hospital in some retired spot for the reception of the inoculated only, and guarding it, in order to cut off all communi-cation between the hospital and the troops, as had been pursued under like cir-cumstances before Boston in the beginning of the war.

pital affairs in proper train at Newark and Hackensack, he left Drs. Foster and Warren in charge, and after taking the necessary steps for providing for the sick at North Castle and Peekskill, he proceeded to join the Commander-in-Chief and the army at White Plains. From thence he followed to the Jerseys, and reached headquarters just before the affair at Trenton. Here he immediately reported to the General, and informed him that by his commission, as well as from former orders received from the Secretary of Congress, "the department at headquarters was under his immediate direction," and that his duties required by the last resolve of Congress having been fulfilled, he had hastened to an attendance upon him without waiting for his commands, at the same time respectfully stating that unless his rank and place were acknowledged and restored to him, he should feel himself obliged to give up his charge. Washington assured him that the difficulties of which he complained were in no way owing to him. "I am here," said the latter, "without any assistance from the Hospital Department, and in case of need I know of nobody to take the direction. It is very strange, and I would have you lay the matter before Congress, in order to have a remedy for this irregularity and inconvenience." In pursuance of this advice, and by the General's permission, he afterwards waited upon that body in Philadelphia for an explanation of the resolves, and for further instructions. Upon presenting himself, Mr. Samuel Adams was deputed to receive him, and gave him to understand that complaints had been brought against him, and that the sufferings of the sick in Jersey, a department which Congress had, as has been just shown, placed under the sole control of another Director, were ascribed to him, with circumstances greatly reflecting on his humanity. These injurious charges Morgan indignantly denied, and entreated to be immediately introduced to the body in order to vindicate himself from them, and have his conduct tested by the strictest scrutiny. This, however, was refused him, that assembly being fully employed in urgent matters, a step which at this day will be the more readily pardoned, when we recollect the critical and alarming state of the country at that time; when, in fact, it seemed doubtful whether the recently assumed independence of the States could much longer be maintained.

Immediately after the interview with Mr. Adams, Congress was forced to retire to Maryland, and there he repeated his application through the same gentleman by letter, to which he never received a reply. Resignation was intimated to him, by some, as the best means of quieting the clamours against him, and freeing himself from further trouble in the burdensome duties of the service, but this he spurned. The dark clouds which hung over the political horizon threatened the total destruction of our cause, and it was, in his opinion, no time for a retreat. He determined to stand by it during the then dangerous period, and prove by his actions the falsity of the charges made against him, and when that had passed, insist on an immediate hearing, or resign.

Accordingly, he a second time repaired to headquarters, but by a resolve of Congress received there, he was again directed to take charge of the sick on the east side of the Hudson. This order he obeyed, and pursued his business as though he had no cause of complaint, hoping for redress as soon as Congress should find leisure, and be in a situation to attend to it. He hastily prepared, however, and on February 1st, 1777, addressed a "Memorial" to the Commander-in-Chief, in which he set forth his manner of conducting the department of the General Hospital, and his actions from the time of his appointment, in the hope of obtaining a Court of Inquiry respecting his conduct; but on the very day after its reception by Washington, orders from Congress were received for his dismission, without assigning any reason for it, and without his having been made acquainted with his supposed offences, or granting an opportunity for the truth or falsity of them to be examined.

Upon his dismissal, the officers of the General Hospital, among whom were John Warren, William Eustis, Philip Turner, and Isaac Ledyard,* names among the best known and brightest in the medical corps of the time, considering themselves in honour obliged to certify facts of which they were acquainted, respecting the conduct and management of the late Director-General of the American army, and Physician-in-Chief, joined in presenting a testimonial to Dr. Morgan, of which the following is an abstract:—

"That the Director-General was ever attentive to any calls made upon him for the supply of the hospital, and for procuring comfortable accommodation, provision, necessaries, and refreshment for the sick. That in particular instances, on

---

* The other surgeons of the General Hospital, Drs. Foster, Adams, McKnight, and Burnet, were on detached service at a distance at the time; but Dr. Morgan states that he is sure, if called on, they would testify to the same faithful discharge of his duty.

10   (73)

the breaking up of regimental hospitals, and routing of the
sick by the approach of the enemy, the sick being poured
upon his care in vast numbers, without either surgeon, mate,
quartermaster, or any one officer to accompany them, Dr.
Morgan having but little assistance for such sudden increase
of numbers, went from house to house to procure quarters
for them, and provision for their support, instructing the
country people in what manner to take charge of them, even
to seeing them provide their meals for present refreshment,
and take care for their future sustenance. That he visited
hospitals, going through each ward in person, to see that the
sick were duly attended and provided for. That he consulted
with the surgeons, and gave his assistance whenever there
was occasion, and showed a readiness to remove any difficulty
that offered to the utmost of his power. That he performed
capital operations himself when present, or assisted the surgeons therein, and stooped to do the duties of a mere mate
in dressing the most simple as well as the most dangerous
wounds of the soldiers in the General Hospital, as an example
and encouragement to the officers under him to attend carefully to that duty. That if the sick have at any time suffered
more than usual, it has been from unavoidable accidents not

in his power to remedy. That there were no complaints of the sick suffering in the General Hospital under their care, but that the uneasiness that arose was concerning the care and provision of those who were retained in their regiments, and not reported to the General Hospital. That they have been witnesses to his constant application and attention to the duties of his station, in which his diligence, assiduity, care of, and humanity towards the sick and wounded were abundantly evident. That the principal causes of the sufferings of the sick the last campaign, proceeded from the regimental sick not being properly reported to the General Hospital, and in some instances to the unavoidable scarcity of particular medicines and other stores, which could not be procured at all times in sufficient quantities, and not, so far as ever came under their notice, to any want of care and attention in the Director-General, but principally from the repeated movements of the army, which exposed the sick and wounded to sufferings that could in no way be remedied, and are firmly of opinion that no person whatever acting in the capacity of Director-General under the same circumstances, could possibly have given universal satisfaction."

The complaints against the Director-General, as before

intimated, arose from his rigid enforcement of the resolves of Congress "that no stores should be issued from the General Hospital to the regimental surgeons," and that "no regimental hospital should be allowed in the neighbourhood of the general one," and his hasty dismission was owing to charges made to Congress in regard to the sufferings of the sick in Canada and Jersey, the direction of which had been taken by their own resolves entirely from his control. That the sick suffered elsewhere, and that there was a real want of many of the material requisites for their relief even in the hospitals under his own eye, he or no other denied. It was indeed unavoidable.* A proper hospital establishment was beyond the abilities of the country. The army, too, was a young one, of which the militia formed a considerable part, was unused to discipline, and exposed to great hardships from a deficiency of proper clothing and stores of all kinds, articles also out of the power of Congress to furnish. Besides this, they were an unsuccessful and retreating army, soured by disappointments and reverses. Officers, as well as soldiers, manifested a reluctance to submit to the necessary discipline of camps,

* *Vide* Appendix II.

and no head of the department whatever, under the same circumstances, could have given universal satisfaction.*

From the time of his entrance into the Hospital Department, Dr. Morgan devoted every hour, and sacrificed every private interest to relieve the sick, and correct abuses. He was not justly chargeable with any neglect, and discharged his duties faithfully and well. Congress, however, was forced from the pressing situation of affairs and the misrepresentations and complaints circulated against him, excited in great measure it is believed by the contrivances of his enemies, to give way to the storm, and removed him unheard, to quiet the clamours of a strong political party.

* The sufferings of the troops in New Jersey from camp fever were very great. Dr. Rush informs us, that of those brought to Philadelphia in open wagons, many perished from hunger and exposure, and that a thousand or more died and were buried in our Potter's Field. In the Northern army, it is stated, that five thousand men had suffered from smallpox, between the 1st of April and 8th of August, at which latter date a General Order was issued prohibiting inoculation; but by February, 1777, the disease had made such head in every quarter, that Washington found it "impossible to keep it from spreading through the whole army in the natural way," and therefore "determined not only to inoculate all the troops of his command, but also all recruits as fast as they came to Philadelphia." (Sparks's Life, vol. ii.)

Thus calumniated and condemned for the faithful discharge of the duties of his office, Dr. Morgan felt keenly the indignities to which he had been subjected, but was supported in his trials by that consciousness of integrity and right conduct which alone can, under such circumstances, give solace. Notwithstanding the injustice which had been done him, he at this time thus wrote to a friend concerning his dismissal: "It is an act into which they were suddenly forced by a party whom political necessity obliged them to gratify. But such is my opinion of the integrity, and such my reliance on the honour of Congress, as to believe that when they are furnished with the materials for judging properly, they will be as ready to do me justice, as a part of them have been to listen to the malice and misrepresentations of my adversaries, and to show their magnanimity, by allowing that they have been capable of an error by their readiness to redress it. I have endeavoured to discharge my duty in what I undertook from principle, according to my degree of knowledge and capacity, with fidelity and diligence; and what I value more than knowledge or capacity alone, with humanity; from whence results the approbation of a good conscience which as my enemies, with all their power cannot give, so neither can they take away."

He continued ardent in the cause of freedom, which he had among the foremost eagerly embraced, but would not rest under the base imputations cast upon his honour, and formally demanded of Congress a Court of Inquiry concerning his whole conduct while Director-General of the Medical Department, and this, though urged on with remarkable energy and perseverance, was from time to time postponed.

At the present day it may seem strange that so just and reasonable a request should long have been denied him, but it will not be so considered when we recall the then existing state of affairs, and the press of other more urgent matters upon Congress.

The country was in the extremest peril, and that body exercising, as it did during the whole course of the war, not only legislative, but also executive and judicial powers, was overburdened with affairs of more consequence than the hearing of complaints, however just. All business coming before it was performed by committees, and from this mode of proceeding, wise and patriotic though they were, yet action on important measures was often dilatory, and it may have been at times even affected by the prejudices or private resentments of some of those forming them.

The Medical Committee of that period was composed, among others, of Gerry, and Samuel and John Adams; men among the most active of the body on other committees, and time and opportunity was really wanting to them to examine into such details as would be necessary for a proper judgment.[*] Thus Morgan, though believed by all who knew him, as well as by popular opinion, and the Commander-in-Chief, to have discharged his high trust with integrity, energy, and marked ability, was yet doomed to be put aside for some more convenient season.[†]

[*] "The whole Congress is taken up almost in different committees, from seven to ten in the morning. From ten to four, or sometimes five, we are in Congress, and from six to ten in committees again. I don't mention this to make you think me a man of importance, because it is not I alone, but the whole Congress is thus employed." (Letter of John Adams to his wife, vol. i. p. 77, 1841.)

During Mr. Adams's term of service in Congress, he was a member of ninety, and chairman of twenty-five committees.

[†] Washington, in his letter to Morgan of January 6th, 1779, gives the most ample testimony of his having discharged his duty with diligence and fidelity, saying, "No fault, I believe, was or ever could be found with the economy of the hospitals during your directorship;" and, moreover, two of the prominent actors in the Medical Department of this period, who have left records on this matter, are of this opinion. One of these, the venerable Dr. Thacher, asserts that the clamours raised against Morgan were unjust, and that no opportunity was

More than two years elapsed before a hearing could be obtained, but justice, though slow, at last came. Congress found time to give ear to his request, and Messrs. Drayton of South Carolina, Harvey of North Carolina, and Witherspoon of New Jersey, were appointed a committee for the purpose of examining into his conduct in the public service, and tracing out the true causes of the sufferings of the sick in the army during the campaign of 1776, and the complaints they produced.

Upon the appointment of this committee, he published a card in the newspapers of the different States, inviting and challenging "every person who had anything to allege against the faithful discharge of his public trust as Director-General and Physician-in-Chief to appear before the above-named gentlemen with evidence in support of their charges, that he might have an opportunity of meeting them face to face to answer their accusations." *

afforded him to vindicate himself from them; and Dr. Rush, who occupied the post of Physician and Surgeon-General of the Middle Department, besides being a member of Congress, and aware of all the proceedings, in a letter to Dr. Bond of the date of February 1st, 1778, says : "In order to avoid the fate of Dr. Morgan, as well as to gratify my own inclinations, I have sent in my own resignation."

* *Vide* Appendix III.

11   (81)

The committee made a full investigation of his conduct, which resulted in an honourable acquittal of the whole of the charges made against him, and their report, after being before the House near three months, "for the perusal and satisfaction of the members, with the evidence upon which it was founded," received from Congress their official sanction, as will be seen in the following proceedings :—

"IN CONGRESS, June 12th, 1779.

"Congress took into consideration the report of the committee to whom was referred the Memorial of Dr. John Morgan, late Director-General and Physician-in-Chief in the General Hospitals of the United States, and thereupon came the following resolution :—

" *Whereas*, by the report of the Medical Committee confirmed by Congress on the 9th of August, 1777, it appears that Dr. John Morgan, late Director-General, and Chief Physician of the General Hospitals of the United States, had been removed from office on the 9th of January, 1777, by reason of the general complaint of persons of all ranks in the army, and the critical state of affairs at that time ; and that the said Dr. John Morgan, requesting inquiry into his conduct, it was thought proper that a Committee of Congress should be appointed for that purpose.

" *And whereas*, on the 18th of September last, such a committee was appointed, before whom the said Dr. John Morgan hath, in the most satisfactory manner, vindicated his conduct in every respect as Director-General and Physician-in-Chief, upon the testimony of the Commander-in-Chief, General Officers, officers in the General Hospital Department, and other officers in the army, showing that the said Director-General did conduct himself ably and faithfully in the discharge of the duties of his office ; Therefore,

" *Resolved*, That Congress are satisfied with the conduct of Dr. John Morgan, while acting Director-General and Physician-in-Chief in the General Hospitals of the United States, and that this resolution be published.

"Extract from the Minutes.

"CHARLES THOMSON, SECRETARY."

To the inquirer of the present day, it appears probable that the difficulties which Dr. Morgan had to contend with in his department, arose primarily from the fact, that in the then state of the Colonies it was impossible to obtain adequate subsistence, shelter, supplies, or transportation for the sick and wounded of the army, and were increased by the imperfect system of organization for the Medical Department which was adopted by Congress. With such a faulty system, it was of course impossible to carry out any plan for the proper treatment of the sick and wounded; and it appears from disputes between himself and other medical officers, concerning rank and precedence, that his office as Director-General and Chief Physician was far from giving him that degree of authority requisite to a successful administration of the affairs

(83)

of his department. Dr. Morgan justly held that a due respect
to rank was necessary and proper; and we find in a letter from
Dr. Brown, a prominent hospital director, that his scrupulous
adherence to a supposed dignity of office gave much offence
to some of the Congress, as well as to his brother officers of
the profession. In fact, in the early part of the war the ad-
justment of rank was a great difficulty in the army, and there
were perpetual disputes concerning it among all classes and
grades. In the medical corps, sub-directors and surgeons
were appointed by Congress for different districts, and the
not unfrequent removal of these officers from one district to
another by order of Congress, was productive of interference
of authority, and jealousies and disputes, very injurious to
the service; and it is not unlikely that Dr. Morgan, though
undoubtedly at times treated with marked want of respect by
some, yet made more complaint to the Medical Committee of
Congress in regard to it than the latter considered mere matters of etiquette required, amid the heavy trials the country
was then passing through. This inference is much strength-
ened, I think, by the following passage in a private letter to
him from Charles Thomson:—

"There is no man, Sir," says the truthful Secretary, "acquainted with you who can doubt your abilities. All the world bears witness of them, and the learned in Europe, who must be allowed to be the best judges, have given ample testimony by the honours they have heaped upon you. While you exercise your great talents for the benefit of those entrusted to your care, your country will honour you, and posterity will do you justice; even though Dr. S——, when you chance to meet, should refuse to give you precedence." In another private letter to him, the Secretary says, "As to rank and precedence, and all that nonsense, there is nothing in the Journals to establish it."

His unjust dismissal from the post of Director-General doomed him to temporary disesteem, and this, to a man like Morgan, highly educated and possessed of great sensibility of character, prostrated him, and from its benumbing influence he never fully recovered. After the publication of his "Vindication" and subsequent honourable acquittal by the Congress of all of the charges which had been brought against him, he withdrew in a great measure from the public eye, and passed his days mostly in retirement and study, which a still sufficient, though shattered estate, allowed him happily the means of indulging in. He continued, however, his services to the

Pennsylvania Hospital, with which he had been long connected as physician, until 1783, when he resigned, as mentioned in its records, "to the grief of the patients, and much against the will of the Managers, who all bore testimony to his abilities, and great usefulness to the institution."* His

---

* The immediate cause of his retiring from the Hospital is a little curious. A custom prevailed in the last century of sending syphilitic patients from the Almshouse to the Pennsylvania Hospital for treatment, as it was then deemed necessary to subject them to a mercurial course carried to salivation, and the accommodations for the purpose were better at the latter than at the former institution. In addition to the expenses of board and nursing, a fee was always charged against the Almshouse by the physician under whose care the case was treated. There is a record on their minutes of two guineas having been paid to Dr. John Morgan. Afterwards, the Managers complained to the Hospital of the charges made to them by Dr. Morgan of seventy shillings for curing each venereal patient. The minutes of the Board of Managers of the Hospital inform us that "the doctor admitted the charge, and refused to relinquish it for persons of that description, thinking it sufficient to attend all other cases gratis; but the Board resolved that there should be no such charge made to the Overseers of the Poor, on which the doctor resigned his place." As stated, this took place in 1783, and the records of the institution mention that "The committee appointed to return the thanks of the Board of Managers to Dr. Morgan for his services to the Hospital, report that they have thanked him, and that the doctor very politely answered 'He was always ready on any extraordinary emergencies to render the institution any further services in his power.'"

health gradually gave way, so much so at last as to induce him to spend a winter in the Southern States, from whence he returned but little benefited.

Of the particulars of Dr. Morgan's life for the two or three years immediately preceding his decease, I have been unable to learn much, though I have had in my possession abundant evidence to show, that in his latter days he was resigned and hopeful, looking forward to another world for that peace which had been refused him here. In a letter to his brother, at Princeton, a few months before his decease, which with 'many others I have had the privilege of examining, he writes, "Wearied with this world, I have for some time past turned my mind more than ordinarily to the thoughts of a better, where I wish to go, but with resignation to the superior will of Him who has the right to appoint the day and hour." He died in the city of his birth, October 15th, 1789, in the 54th year of his age, and was buried in the ground attached to St. Peter's Church. In the records of the day, there is a bare mention of his death and age, and no stone remains at the present time to point out his last resting-place. So far as I have been able to ascertain, the only notice of him is to be found in the short but grateful tribute which Rush, his friend and successor in the Chair of the University, paid his services, and the brief mention of him furnished by Thacher, mostly drawn from it.

His wife, Mary, one of the daughters of Thomas Hopkinson, Esq., to whom he was married on his return from Europe in 1765, preceded him to the tomb, having died in 1785. She was a sprightly, agreeable woman, and they lived most happily together. He left no issue, and his property, which was considerable, he bequeathed to his brother, Col. Morgan, whose descendants now reside in the city of Pittsburgh. In person, Dr. Morgan was under the medium height, was delicately made, with an expressive and handsome face. He was polite and gentlemanly in his manners and address, but prolix in conversation.

A fine collection of paintings and engravings which he had made in Europe, together with a choice and valuable library, and his manuscripts, the labour of ten years, were all either destroyed by the enemy at Bordentown, New Jersey, whither he had removed them from Philadelphia for safety, or were consumed by fire at Danbury, Connecticut, in the destruction of that place by the troops under Governor Tryon.

His literary productions, besides his Thesis and his Discourse upon the "Introduction of Medical Schools in America," published in 1765, to which we shall have occasion again to revert, were, "A Recommendation of Inoculation according to Baron Dimsdale's Method," printed in 1776, and a work written some years prior to the Revolution upon "The Reciprocal Advantages of a Perpetual Union between Great Britain and her American Colonies." In 1777 he published "A Vindication of his Public Character in the Station of Director-General of the Military Hospitals." In addition to these, he contributed to the Transactions of the American Philosophical Society for 1786, "An Account of a Pye Negro Girl and Mulatto Boy;" an article "On the Art of Making Anatomical Preparations by Corrosion," an abstract from the essay presented by him to the French Academy of Surgery; and an interesting paper "On a Snake in a Horse's Eye, and of other Unnatural Productions of Animals."

The life of Morgan affords a bright example of acquirement, perseverance, usefulness, and a noble love and devotion to the profession of his choice. The title of founder of Public Medical Instruction in America justly belongs to him, yet no memorial exists to recall his great services to his profession and his Alma Mater. His very name is almost

12

forgotten by the mass of the brotherhood; and even here, in the city of his birth and labours, how few are there who are aware of the benefits he has conferred upon us! The school originated by him still flourishes, receiving, as he himself foretold, "a constant accession of strength, and annually exerting new vigour, and has given birth to numerous other useful institutions of a similar kind, spreading the light of medical knowledge through the whole American continent."*

His Discourse on the "Institution of Medical Schools in America" is, considering the state of medicine at the time it was written, a remarkable production, and should be republished and circulated as an act of justice to his memory. Although the science has advanced immeasurably since that day, his enlarged views of what is required of a medical practitioner by preliminary education, his high-toned sentiments

---

* The above was written in 1846. Since that time a portrait of Dr. Morgan has been placed in the museum of the University. It is a copy from a fine painting in my possession, after the original by Angelica Kauffman. Writing in 1827, the late Dr. Meigs eloquently and justly says: "He who in Greece or Rome would have had statues of brass and marble voted to his memory has in the one-third part of a century gone out of remembrance so completely that of the many of the hundreds who partake of the benefits of his school, very few have ever heard of his name."

DR. HABRAHAM CHOVET, ÆT. 80.

DR. VAN EOKHOUT.

regarding its practice, honours, and emoluments, his recommendations of clinical teaching and hospital instruction, his
recital of the years of labour spent by him in preparation for
its active duties, in addition to its historical value, all make
this now very rare tract worthy of such attention.

In the year 1774 a foreign physician took up his residence
among us who, besides becoming distinguished as a practitioner, laboured zealously in teaching practical anatomy,
and is deserving of honourable mention from having aided
materially in its progress.

I allude to **Dr. Abraham Chovet.** He was a native of
England, and had devoted much of the early part of his life
to the study of anatomy under the ablest teachers of Europe,
and lived first at Barbadoes, and afterwards in the Island of
Jamaica, from whence he came to this place. A well-known
antiquarian has thus described him from recollection, as he
appeared a few years previous to his decease. "This aged
physician was almost daily to be seen pushing his way, in
spite of his feebleness, in a kind of hasty walk or rather
shuffle, his head and straight white hair bowed and hanging
forward beyond the cape of his black old-fashioned coat,
mounted by a small cocked hat, closely turned upwards

upon the crown behind, but projectingly and out of all proportion cocked before, and seemingly the impelling cause of
his anxious forward movements; his lips, closely compressed
(*sans* teeth) together, were in constant motion, as though
he were munching something all the time; his golden-headed
Indian cane, not used for his support, but dangling by a
black silken string from his wrist; the ferule of his cane and
the heels of his capacious shoes, well lined in winter time
with thick woolen cloth, might be heard jingling and scrap-
ing the pavement at every step; he seemed on the street
always as one hastening as fast as his aged limbs would permit him to some patient dangerously ill, without looking at
any one passing him to the right or left."

The doctor was an eccentric character, full of anecdote and
knowledge; and, tradition informs us, possessed of great sar-
castic wit. He was much in the habit of using certain
expletives in his ordinary conversation, which, in the opinion
of those who best knew and appreciated him, were thought
to be neither useful nor ornamental. An anecdote, strikingly
illustrative of these points of his character, has been mentioned to me as being well known to those of former days.
The doctor happened to be overtaken at the house of a mem-

ber of the Society of Friends by a heavy shower of rain, and as he insisted on pursuing his way during its continuance, the friend kindly offered to loan him his overcoat, adding the condition that he was not to use hard words while it was upon his back, to which the latter assented.   On returning the coat he was asked, "Well, doctor, didst thou swear whilst thou hadst on my coat?" "No," replied he, "but there was a damnable disposition to lie."   The doctor was a well-known Tory, and from his advanced years, and his being English born, no offence was taken at his expressing his political opinions freely, which he did on all occasions, and never lost an opportunity of joking and quizzing his friends, of whom he had numbers of the opposite party.   On one occasion, being sent for to visit the Spanish Minister, M. Mirailles, the ambassador ordered his carriage to convey him home; the doctor full of fun, and delighted at the opportunity for a laugh which it afforded him, directed the coachman to drive slowly by the Coffee House, it being an hour when he knew the merchants would be all congregated there.   The equipage of the Don, as Minister from a friendly power, was, at that time of high political excitement, well known, and when perceived to be advancing, the merchants drew up in order, hats off, to pay their respects to him. The doctor kept himself close back in the carriage until directly opposite the building, the gentlemen all politely bowing, when he suddenly popped out his head, with "Good morning, gentlemen, good morning, I hope you are all well; thank you in the name of his Majesty King George," and drove off, laughing heartily at having quizzed the Philadelphia Whigs. At the commencement of the war, however, he was looked on as a "dangerous man," and only escaped being carted through our streets along with Dr. Kearsley by secreting himself in the stable of Mr. Marshall, and in May, 1777, was forced to take the oath of allegiance to Congress.

Dr. Chovet brought with him to Philadelphia a complete and beautiful collection of anatomical preparations in wax, which he had made in Barbadoes in 1744, and gave courses of anatomical and physiological lectures. The introductory to his first course in November, 1774, was delivered in great form, being attended by Governor Thomas Penn, Rittenhouse, Dickinson, the clergy, physicians, and other of the most influential men of the city. It consisted of a Latin oration on the origin and dignity of physic, and was followed by a learned discourse in English on the history and progress of the sciences of anatomy and physiology. These lectures were illustrated by his wax models, together with dried preparations and injections, and were given at a hall in Videll's Alley, Second Street. The building still stands upon the south side of the alley. It is a quaint-looking old-fashioned two-storied brick house, with a steep pitched roof and dormer windows; it is now used as a carpenter shop. In this locality his lectures continued to be given, and his anatomical collection remained until 1777, when the latter was removed to his dwelling in Water Street, near the old ferry. Here he erected an amphitheatre in which his lectures were afterwards delivered, the first being given there in January, 1778. John Adams, in his Diary (Works, vol. ii.), speaks of his visiting the museum of Dr. Chovet, when in Philadelphia, and records that his cabinet was much "more exquisite than that of Dr. Shippen at the Hospital." The doctor, he added, reads lectures for two half joes a course, which takes up four months. Dr. Chovet died March 24th, 1790, aged eighty-six, and was interred at Christ Church. He visited his patients in all weathers on foot until within a few weeks of his death; his faculties having exhibited no marks of decay, and finally being carried off by some acute disease. We are told by

Dr. Rush that he was so sensitive to cold, that he slept "in a large night-gown, under eight blankets and a coverlet, in a stove room, for many years before he died." He applied his wit to his years, and used to say that "that physician was an impostor who did not live until he was eighty." He made it his dying request, that he might have a plain funeral, and that no bell might be tolled on the occasion, as he did not wish to disturb sick people by such unnecessary noise. His daughter, who died in 1813, bequeathed to the Pennsylvania Hospital a portrait of her father, painted by Pine, which now hangs in that building; his anatomical cabinet having been previously purchased of her by that institution for an annuity of £30, payable out of the medical fund. Dr. Coste, the chief medical officer of Rochambeau's army, in a tract which he published at Leyden, in 1784, speaks of Chovet as "a man skilled in all things pertaining to medicine, and especially in anatomy and surgery;" and the Marquis de Chastellux,* who was on intimate terms with Dr. Chovet when in this country in 1780, says that "many of his wax preparations were equal to those of Bologna." This traveller thus speaks

---

* Travels in North America, vol. i. p. 233. London, 1787.

of the doctor himself: "He is a perfect original. A man of application, and of great natural vivacity; his reigning taste is disputation. When the English were at Philadelphia he was a Whig, but before and since they left a Tory. He is always sighing after Europe without resolving to return, and declaiming against the Americans he still remains among them. His design in coming to the continent was to recover his health, so as to enable him to cross the seas; this was at the commencement of the war, and since that time he imagines he is not at liberty to go, though nobody prevents him."

Another physician of note who, towards the close of the Revolution, made Philadelphia his residence, was **John Jones.** Dr. Jones was a native of Long Island, and after completing his medical studies in Philadelphia, improved himself still further by a visit to Europe, and, upon his return, settled in New York, devoting himself particularly to the practice of surgery. He was the first to perform the operation of lithotomy in that city, and his success was such in several cases which soon presented themselves to him that he became well known as an operator throughout the Middle and Eastern States. Upon the foundation of the New York Medical School, he was appointed to the professorship of

13   (97)

surgery. At the commencement of our war, he published his "Plain Remarks upon Wounds and Fractures," a work intended principally as a guide to young surgeons of our army in the classes of accidents to which their attention then was continually directed. This book, which embodies the sentiments of the best surgeons of the period on the subjects treated of, with the result of the author's own observations, contains much valuable matter, and is well put together. It passed through three editions. The first was published in New York in 1775; and the two latter at Philadelphia in 1776 and 1795. He removed to Philadelphia in 1779, and was in the following year elected one of the surgeons to our hospital; and upon the foundation of the College of Physicians, in which he took a prominent part, was made one of its Vice-Presidents. He died in 1791, aged sixty-two.

Though not professional men, yet as well on account of their devotion to the kindred sciences of botany and natural history, as from having aided the progress of medicine in Pennsylvania, it would be improper in any sketch such as this to pass over altogether in silence the names of James Logan and John Bartram.

James Logan was a man of learning and strong abilities, who assisted materially in encouraging medical science among us, as well by the formation of a library rich in the most valuable and rare works relating to it and kindred subjects, as in the countenance afforded by him to anatomical pursuits. It was under his auspices, and in a building belonging to him,* that Dr. Cadwalader first made his anatomical demonstrations —a use, so strong in those days was the feeling against dissections, to which few would have been found willing to appropriate their property. His translations of Cicero "On Old Age," and Cato's "Distichs," were among the first translations from the classics made on this Continent. Besides these, he was the author of a scientific work entitled, "Experimenta et Meletemata de Plantarum Generatione;" or, Experiments on Indian Corn, with his observations arising therefrom on the Generation of Plants. This was published at Leyden in Latin in 1739, and was afterwards, in 1747, republished in London, with an English version on the opposite page, by Dr. Fothergill.

Of **John Bartram**, it is sufficient here to say that he

* On Second Street above Walnut, on the site afterwards occupied by the Bank of Pennsylvania.

has been justly styled one of the fathers of natural history in North America, and that in his specialty he was preëminent, and was pronounced by Linnæus to be the greatest practical botanist whom the world had seen. He established on the banks of the Schuylkill, near Philadelphia, the first botanical garden in America, corresponded with many of the distinguished philosophers of his time, was elected a Fellow of the Royal Society, as well as of several other scientific associations of Europe, and was made American Botanist to King George the Third, which appointment he held until his death in 1777. In 1751 he published an edition of Dr. Short's "Treatise on Plants," with an appendix containing a description of the medicinal properties of those peculiar to America, which is, I believe, the first attempt at the formation of an American Materia Medica. He performed many journeys in the pursuit of his favorite study, and published "Observations on the Inhabitants, Climate, Soil, etc., made in his travels from Pennsylvania to Onondaga." At the age of seventy he travelled through East Florida, in order to explore its natural productions, and afterwards published a journal of his observations. He died on the 22d of September, 1777, aged seventy-eight years.

PENNSYLVANIA was the first of the Provinces to adopt a system of regulations for the protection of the community against sickly vessels, as well as for the erection of a hospital to prevent the spread of contagious diseases. The earliest epidemic of which any account has come down to us after the settlement of Philadelphia was in 1699, when yellow fever prevailed, and proved exceedingly fatal— "six, seven, and sometimes eight dying in a day, for several weeks together, and few, if any, houses being free of the sickness." The disease was believed to have been introduced here, and its occurrence led in the following year to the passage of an Act by the Assembly "to prevent sickly vessels suddenly coming to this port." How far the measures then adopted were carried into effect we have no means of judging, for from this date till 1720 there is nothing noted in regard to sanitary measures, though it may be inferred from the

(101)

subjoined statement of Dr. Græme that the laws then enacted were enforced. In this latter year it is mentioned that Patrick Baird, Chirurgeon, was appointed Port Physician.

In 1738 two vessels with passengers arrived with the "Palatine Fever," and created so much alarm that the Governor, in his message to the Assembly, informing them of the fact, adds "that the law of 1700 to prevent sickly vessels from coming into this government has been strictly put in execution."*

A few months after this occurrence, it is recorded in the minutes of the House that a petition from Dr. Græme was presented, setting forth "that by order of several governors *for upwards of twenty years past*, he has served the public by visiting and reporting the state of sickly vessels arriving here, to the apparent risque of his own health and life, for which, with other services done by him for the public,† he has never yet received any reward, and praying the house to take the same into consideration." The sum of £100 (currency) was

* The masters were compelled to land the sick "at a convenient distance from the city, and to convey them, at their own expense, to houses in the country proper for their reception."

† "Going a journey about Indian affairs."

voted to him.* In June, 1741, the doctor desired to be excused from further service, and was succeeded by Dr. Lloyd Zachary, the Assembly at the time of his appointment directing that "he be paid a reasonable reward for services which he shall do in visiting the said vessels."

As early as 1738, Governor Thomas had recommended to the Assembly the erection of a pest house or hospital for infectious diseases, but the proposition was not assented to by them till three years afterwards, when the subject being again urged, it was agreed to, and Fisher's Island, situated at the junction of the Schuylkill and Delaware rivers, was purchased for that purpose, and a hospital erected thereon.†

The vessel that brought William Penn, with about one hundred passengers, to our shores in 1682, suffered during the voyage from smallpox, which proved fatal to thirty of them, though it does not appear to have been communicated to the settlers who had preceded him. In the year 1701, the disease prevailed here, and was "general and mortal."

* From the votes of the Assembly, we find that his services were remunerated at the rate of a pistole a visit.

† An interesting historical sketch of our quarantine by the late Wilson Jewell, M.D., was published in 1857, where this subject will be found fully treated.

In 1726 a vessel with the smallpox on board arrived, though nothing was said of its spreading in the town. In 1730 a severe epidemic of it prevailed, and it was at this period that inoculation was introduced among us.

Inoculation.—This practice had been brought into England in 1721, and adopted to some extent in the sister colony of Massachusetts in the same year, and the prevalence of the disease led a number of our citizens to submit themselves to it.

Under the date of March 4th, 1730, the *Pennsylvania Gazette* announces that "Joseph Growden, Esq., the first patient of note that led the way in inoculation, is now upon the recovery, having had none but the most favorable symptoms during the whole course of the distemper, which is mentioned to show how groundless all those extravagant reports are that have been spread through the Province to the contrary."

One of the earliest advocates for inoculation, Franklin, who well knew how to influence men by making appeals to their pockets, thus notices the disease in his paper of July 8th, 1731: "The smallpox has quite left the city, the number of those that died here of that distemper is exactly 288, and no more; sixty-four of the number were negroes; if these may be valued one with another at £30 per head, the loss to the city in that article is near £2000." Nevertheless, inoculation met with much opposition. Between the autumn of 1736 and spring of 1737 smallpox was very rife, and "proved as mortal in the common way of infection as was ever known in these parts." It is reported that during its continuance only one hundred and twenty-nine persons underwent inoculation, of whom but one, an infant, died; and that of the number inoculated "one was in the fifth or sixth month of her pregnancy, notwithstanding which she did well." Even Franklin at this time must have become doubtful of its benefits, for from his autobiography we learn that in this epidemic he lost a fine boy four years old by the disease taken in the common way, and that he afterwards greatly regretted not having given it to him by inoculation. Here, as elsewhere, the people were divided as to the propriety of inoculation, some contending warmly for its introduction, and others as strongly opposing it, looking upon the practice of "soliciting a distemper before nature was disposed to receive it as a tempting of Providence and a suggestion of the enemy of all righteousness," and asserting that the surgeons concealed or diminished the true number of deaths occasioned by it, at the same time that

14 (105)

"they magnified the number of those who died of the disease in the common way."

In 1750, the subjects of smallpox and inoculation still excited much attention in our community, and Dr. Adam Thompson published a tract "On the Preparation of the Body for the Smallpox," in which it was asserted that inoculation "was so unsuccessful in Philadelphia, that many were disposed to abandon it." This work of twenty-four quarto pages I have been unable to obtain. It is spoken of as having merit, "being written in a modest and plain style, the arguments made use of as highly plausible, and the author as actuated with a generous desire to communicate salutary advice in the management of a distemper which has proved fatal to multitudes." In it a cooling regimen is recommended, and upon the suggestion of Boerhaave, he states, "that he was led to prepare his patients for the infection by a composition of mercury and antimony, and that he had employed it for twelve years with great success." The production was severely attacked, among others by Dr. Kearsley, who, in the following year, put forth "Remarks on a Discourse on Preparing for the Smallpox," which led to a rejoinder from the well-known Dr. Alexander Hamilton, of

Annapolis, Maryland, entitled "A Defence of Dr. Thompson's Discourse," and was published here by Bradford in the same year.

In the year 1756 the smallpox was again prevalent, and, as we learn from Dr. Hamilton, of Bush Hill, "raged terribly all the summer and autumn, and swept away numbers of people."* Some British troops under the command of Col. Boquet, who arrived in Philadelphia about this time, increased the ravages of the infection, so much so that Governor Denny, in his message to the Assembly in the month of December, said: "The smallpox is increasing among the soldiers to such a degree that the whole town will soon become a hospital." Inoculation was practised to some extent, but was not general, and the great prejudice against it was only slowly overcome. At this period a tract appeared from the pen of Dr. Laughlin Macleane,† entitled

---

* Letter to Dr. Hill.

† Dr. Macleane was an Irishman by birth, who had been a collegiate acquaintance of Goldsmith, graduated in medicine at Edinburgh, and came out to America, a young man, as surgeon to some British troops. He resided in Philadelphia for a number of years, and has made some noise in the world from his name having been mentioned in connection with the authorship of "Junius."

"An Essay on the Expediency of Inoculation, and the Seasons
most proper for it. Humbly inscribed to the inhabitants

He must have left the army soon after coming here, for he carried on business as
a druggist in Second Street near Market, at the sign of the "Golden Pestle," in
partnership with one Stewart, but I believe never practised medicine. In 1761
he proposed a scheme for the erection, in the neighbourhood of the city, of warm
and cold baths, connected with a public garden and a house of entertainment, by
a subscription lottery. The project, however, was considered unfriendly to
morals, and petitions were addressed to the Governor by a number of prominent
citizens, as well as by societies, to prevent the scheme being carried into effect.
A part of one of the petitions is very curious, and I think may amuse the reader.
In this it is said, "That they believed that a public ground will be a nursery of
all kinds of dissipation. How destructive such places are to the morals of a
people, what they usually terminate in, and how ill-suited they are to the circumstances of this young city, and the former character of its inhabitants, we
need not mention to your Honour. Were there nothing more in view than what
is pretended, it might be effected with as near a few hundred pounds as there are
thousands proposed. Were a hot or cold bath necessary for the health of the
inhabitants of the city they might, at a small expense, be added to the hospital
[hot, cold, and steam-baths had been introduced into this institution by Dr.
Bond, soon after its foundation], put under the sober government of that place,
and kept separate from those used by the patients, and as to a public place for
walking, the State House green or garden is, by a law of this Province, set apart
for that purpose."

The enterprise was abandoned, and the doctor soon after left the country. Mr.
Graydon informs us that "he was considered to have great skill in his profession,

(108)

of Philadelphia. Printed by William Bradford, at the corner house of Market and Front Streets." The style of this little work is quaint. In it numerous extracts from medical writings are given, with classic quotations from Greek and Latin authors. The chief argument against inoculation by scrupulous persons, the author tells us, was from conscience, "they deeming it presumption to tempt the Almighty by inflicting

as well as to be a man of wit and general information, but that he had never known a person who had a more distressing impediment in his speech." In 1767 he became Under Secretary to Lord Shelburne, and afterwards, notwithstanding his misfortune in speech, got a seat in the House of Commons. Lord North, after this, conferred upon him the Collectorship of the Port of Philadelphia, when he came ont a second time, but returned to Eugland the following year, and got an appointment in India as a sort of confidential agent to Warren Hastings. In 1776 he went to England as agent in London to Mr. Hastings, who placed in his hands his resignation as Governor-General, instructing him that it was not to be handed in unless it was "ascertained that the feeling at the India Board was adverse to the Governor-General." Circumstances afterwards occurred when Macleane thought himself justified in producing the resignation with which he had been intrusted. Mr. Hastings denied that his agent had acted in conformity with his instructions, and what they had been he owned he had forgotten, and had no copy of them, though the fact was attested by several, in whose presence the orders were communicated to Macleane. He perished on his return to India in 1777, the vessel in which he embarked never being heard of after she quitted the Cape of Good Hope.

distempers without his permission." After ably combating this opinion, he closes this part of his subject by observing, "Much of this nature are the objections made to the use of Mr. Franklin's invention for defending us from the fatal effects of lightning, certainly one of the most signal benefits to mankind," and argues that the application of the means thus offered is not presumption, but a command to endeavour to avail ourselves of the means of safety which God has left to our sagacity, which parallel reasoning offers an argument equally strong in favour of inoculation." His remarks on the preparation of the body for the disease and its treatment are excellent, and are not surpassed by the best writers of the time. He condemns strongly the custom of using an universal preparative, as was then countenanced, even by some of the Faculty, or, as he expressed it, "the preposterous method of preparing all their patients after one and the same manner, as most offensive to common sense." "The indiscriminate use of mercury," the most common mode, he disapproves of. He also objects to the too free use of cordials and spirits in the treatment, and more particularly "the infernal practice of blistering by rote, whether there be an indication for it or not," recommending in their stead the

(110)

judicious practice of Sydenham, Mead, and Huxham, with whose works he seems quite familiar.

In passing, I would refer to a practice much resorted to a few years since in this neighbourhood, in scarlet fever, and perhaps still popular with some practitioners, viz., that a slice of bacon for the throat "is mentioned by him as a favourite treatment with nurses and the lower people" in smallpox at that period.

Dr. Macleane, it may be here mentioned, had been a pupil of Doctor Rutherford, of Edinburgh, in 1753, and in the course of his little work takes occasion to pay a grateful tribute to him. As this gentleman had been the instructor of most of the ancient physicians of this city, who resorted to Europe for their medical education, and is often mentioned by them, I have thought it might prove of interest to those of the present day to extract it, as showing something of the mode of teaching at that time in Scotland. "A tribute due to his worth from his pupils, who can never sufficiently acknowledge the advantages they have reaped from his labours; above all, from his excellent institution of clinical lectures, where they daily saw him put in practice on num- berless patients the salutary precepts which he had before

(111)

taught them in his class. If great abilities constitute, if a tender heart and extensive charity adorn the real physician, no man ever deserved the title better; no man graced the science more."

Inoculation, however, did not make that progress among the people which was looked for, and Dr. Franklin, at that time in London, who was now a warm upholder of the practice, believing that the expense of the operation, which he says "was pretty high in some parts of America,"* might have been in the way of its adoption, judged that a pamphlet written by a skilful practitioner, showing what preparation should be used before the inoculation of children, and the precautions necessary to avoid giving the infection at the same time in the common way, how the operation was to be performed, and "on the appearance of what symptoms a physician was to be called," might be a means of removing that objection of expense, render its adoption more general, and thereby save the lives of thousands, "prevailed upon Dr. William Heberden to write some account of the success of Inoculation, and Plain Instructions for the same," and that

* Dr. Potts informs us that in Philadelphia they seldom charged less than three pounds.

gentleman generously, at his own expense, printed a very large impression of the work, which was distributed in America. It was handsomely issued in a quarto form in 1759, was largely circulated in and about Philadelphia, and attracted much attention. In this work the operation is recommended at any season of the year, and at all ages, except in the very young and old. Merely weakly constitutions, and those tainted by some hereditary distemper, were not discouraged from being inoculated, but "on breeding women, no consideration whatever should tempt us to perform it, unless we can suppose an absolute certainty of their catching the disease in the common way." In the same year in which this pamphlet appeared, Dr. William Barnet, of Elizabethtown, New Jersey, a gentleman very experienced in the matter, was invited to Philadelphia to inoculate for the smallpox, and opened a house for that purpose, the first private hospital of the kind in Pennsylvania, of which I find any mention. At this time, Dr. Redman, too, published "A Defence of Inoculation," recommending the practice to his fellow-citizens in the most affectionate language, which tended to bring it much more into esteem.

In January, 1773, smallpox again prevailed, and Dr. Glentworth opened a hospital for inoculation. In the winter of 1774 it was also prevalent, and great alarm was created from the fact that no less than three hundred, out of thirteen hundred and forty-four deaths which occurred in the city and liberties during the year, were occasioned by that disorder in the natural way. The chief of these were children of poor people, who could not afford the expense of inoculation, and were unable to procure proper persons to perform it. To remedy this "a society for inoculating the poor" was established, and eight of the principal physicians of the day volunteered to perform the operation, prepare them for it, and also to attend them at their own houses, free of expense. This was done extensively till the month of September, 1774, when the physicians of the city met together and agreed to inoculate no patients during the sitting of Congress "as several of the Northern and Southern delegates are understood not to have had that disorder." In 1776 Morgan tells us "the practice was very common in the Middle States," and it continued to be so till the introduction of vaccination. This took place in 1803, when a printed address was circulated, signed by fifty-seven practising physicians, headed by the venerable Redman, recommending vaccination, and very soon children generally were submitted to it.

MEDICAL SOCIETIES.

A medical society was instituted in Philadelphia on the 4th of February, 1765, and was, as far as I can ascertain, the first professional organization formed in the Colonies.* It was called "The Philadelphia Medical Society." This continued in operation till the 11th of November, 1768, when it was united with the "American Society for Promoting Useful Knowledge,"† it being judged "that the union might be beneficial to the community, and the ends proposed of both be thereby better answered." Accordingly, at the meeting of the latter society, held on the 11th of that month, a number of the members of the Medical Society attended,‡ and the Medical Fellows were appointed a standing committee

* The first Medical Association in New Jersey was held at New Brunswick, in July, 1766, and the Massachusetts Medical Society, which is generally stated to have been the first attempt at professional organization in America, was not originated till 1781.

† Now known as the American Philosophical Society.

‡ The members at the time of union consisted of Drs. Græme, Cadwalader, Redman, Morgan, Kearsley, Clarkson, Bayard, Harris, Rush, Souman, Glentworth, and Potts.

"to consider and report upon matters relative to physic."
By this society a number of dissertations on medical subjects
were received, and were published in octavo form, as well as
in the newspapers—at that period the chief medium which
the country afforded for diffusing such information. Several
of these publications in the book form are now in my possession. Among them are "An Essay on the Virtues and Uses
of several substances in Medicine, that are the native growth
of America;" A Dissertation on the Causes, Nature, and
Treatment of Apoplexy;" "On the Dry Belly Ache, or Ner-
vous Cholic;" On Catarrhal Peripneumony;" and "On Con-
sumption." A second medical organization, termed the
"American Medical Society," was founded in 1770, by a
number of students who had assembled in this city "to hear
the lectures of the medical professors, and who judged they
might derive advantage from associating themselves in order
to discuss various questions in the healing art, and to communicate their observations on different subjects." The
society consisted of senior and junior members, and soon
ranked among its active seniors many of the most eminent
characters in our city. Its meetings were held weekly during
the continuance of the medical lectures. Dr. Shippen was

its President in 1790. It continued in operation till November, 1792, and reports of cases read before it are in print.

From the correspondence of Mr. Thomas Penn with the Secretary of the Province, it appears that as early as the year 1767 Dr. John Morgan was actively engaged in an effort to establish "a College of Physicians;" but in this he was frustrated by the Proprietaries refusing to grant a charter for the purpose, looking upon it as "too early for such an establishment," and the unsuccessful effort thus made prevented any scheme of the sort being carried into execution till after the war of Independence.* Towards the close of the year 1786, the physicians of Philadelphia, influenced by the conviction of the many advantages that have arisen in every country from literary institutions, associated themselves under the name and title of "The College of Physicians of Philadelphia." The first stated meeting of this body was held on the 2d of January, 1787, and two years afterwards it secured its

* Mr. Penn's letter is as follows (extract of a letter from Thos. Penn to Richard Peter, dated Feb. 27th, 1767): "I have had a letter from Dr. Morgan, and proposals for erecting a College of Physicians. I think it very early for such an establishment, and wish the faculty would not press for such a thing. I shall confer with Dr. Fothergill upon it."

Act of Incorporation. The objects of the association, as expressed first by its constitution, and afterwards in the preamble of the charter, are "to advance the science of medicine, and thereby to lessen human misery, by investigating the diseases and remedies which are peculiar to our country; by observing the effects of different seasons, climates, and situations upon the human body; by recording the changes that are produced in diseases by the progress of agriculture, arts, population, and manners; by searching for medicines in our woods, waters, and the bowels of the earth; by enlarging our avenues to knowledge from the discoveries and publications of foreign countries; by appointing stated times for literary intercourse and communications; and by cultivating order and uniformity in the practice of physic." The following were the original fellows who composed it, viz: John Redman, John Jones, William Shippen, Jr., Benjamin Rush, Samuel Duffield, James Hutchinson, Abraham Chovet, John Morgan, Adam Kuhn, Gerardus Clarkson, Thomas Parke, and George Glentworth.

No law has ever existed in Philadelphia regulating the practice of medicine or establishing a medical police; but on the occurrence of violent epidemics, and upon questions of medical jurisprudence, the college has been consulted by the civil authorities both of our State and City.

From the institution of the college, one of their principal views was the formation of an American Pharmacopœia. To make this work useful to the whole country, a circular letter was addressed by them, in the year 1789, to all the known medical societies, as well as to many eminent practitioners in the United States, requesting their advice and assistance; but the general apathy on the subject, and the small number of communications received by them, retarded the completion of their design.*

In 1793 the college published a volume of "Transactions," containing an Address to the members on its establishment, in which is set forth the object of the institution, and suggesting the many resources which our country offers for the improvement of medicine; as well as contributions from

---

* A small collection of recipes intended for the use of the Revolutionary army surgeons was published by Dr. William Brown in 1782. Copies of this, and of the circular letter of the College, both of which are exceedingly rare, are in my possession. The first work of the kind accomplished in the United States was by the Massachusetts Medical Society in 1806, but no National Pharmacopœia was published until 1820.

Rush, Currie, Jones, and others. In 1798 they published a volume entitled "Facts and Observations relative to the Nature and Origin of the Pestilential Fever which prevailed in this City in 1793, 1797, and 1798;" and in the year 1806, another volume of "Additional Facts and Observations relative to the Nature and Origin of the Pestilential Fever."

After the date last mentioned, their publications were discontinued until 1841, when they were recommenced, and since that period have been regularly issued. In 1858 the mode of their publication was changed; they having from that date appeared in the "American Journal of the Medical Sciences."

In addition to the medical societies mentioned, two others were also incorporated by our Legislature, and were in active operation before the close of the century, "The Philadelphia Medical Society," instituted in 1789, and the "Philadelphia Academy of Medicine," of which Dr. Physick was the first President, instituted in 1797, with the particular object of inquiring into and elucidating the nature of pestilential diseases. No State Society existed in Pennsylvania until the year 1848.

The Early History of Medicine in Philadelphia.

PRIVATE AND SUMMER COURSES OF LECTURES.

Private and summer courses of lectures were begun in
Philadelphia at an early date, and did much to attract
students to our city.

Those of Shippen upon Anatomy and Midwifery, have
already been particularly referred to.   In 1766 Dr. Bond commenced Clinical Lectures at the Hospital upon the Practice
of Medicine as well as upon Midwifery.   Dr. Rush lectured
upon Chemistry in 1774, and Dr. Chovet upon Anatomy in
the same year.   All of these attracted considerable notice,
and the courses of the latter gentleman were annually continued for ten or twelve seasons.

In 1784 Dr. John Foulke lectured upon Anatomy and Surgery, and opened an Anatomical Hall, "with a determination
to put the character of a Philadelphia anatomist upon a higher
footing than it had ever before been;" at the same time he
took care to assure his fellow-citizens "that in his pursuit he
was determined to observe every attention to decency, solem-
nity, and punctuality."   The fee demanded by him was twelve
dollars.   His lectures and anatomical rooms were kept up till
his death in 1796.   Dr. Foulke had graduated at the College
of Philadelphia in 1780, and afterwards perfected himself in

the branches taught by him in Europe. He faithfully fulfilled the promises he made at starting, and proved an able, successful, and eloquent teacher. His hall was well patronized, and he did much to promote the study of practical anatomy among us.

In 1789, and subsequent years, I find "Dr. J. H. Gibbons, of Arch Street," advertising his lectures on the Theory and Practice of Medicine.*

In addition to these, Dr. Benjamin Duffield, at an early period, 1793, commenced summer lectures on Midwifery, which were continued till his death in 1799. He likewise, for several seasons, lectured upon Diseases of Hospitals and Jails, "and the American Practice of Physic." He was succeeded in his obstetric lectures by Drs. Church and James. Towards the close of the century Dr. Price, of London, lectured upon the Theory and Practice of Physic, as well as upon Midwifery and Diseases of Women and Children, and at about the same period, 1797, Dr. Dewees presented himself before the public as a teacher "in a regular and extensive course on Obstetrics."

* This gentleman was a Pennsylvanian, and graduated at Edinburgh in 1786. He died October 5th, 1795, ætat 36.

MEDICAL PUBLICATIONS AND LIBRARIES.

In colonial times books were expensive luxuries, yet many of the professional men mentioned in the foregoing pages possessed choice collections of the standard works of the day, and the best editions of the older authors, and for a long period previous to the Revolution there existed in our city a public library, the Loganian, affording ample means for acquiring all that was known upon anatomy, surgery, and the kindred sciences in the English as well as in foreign languages.

As might be expected, the original productions published here were few, the only ones issued in the last century, of which I am aware, being the tracts of Cadwalader, Thompson, Kearsley, Hamilton, Redman, and Macleane, and the works of Morgan, Jones, Rush, Currie, Cathrall, and Deveze, which have already been mentioned. The reprints of English works up to the period of the war embraced about an equal number of volumes. The first of these was an edition of "Short's Medicina Brittanica," with notes showing the places where many of the plants are to be found in these parts of America, their differences in name, appearance, and virtue, from those of the same kind in Europe, and an appendix containing a description of a number of plants peculiar to America, their uses, virtues, etc., edited by the well-known botanist, John Bartram. It was printed by Franklin in 1751. Two editions, the first of which was in quarto form, of the lectures of Cullen on Materia Medica appeared in 1775 and 1789; his work on the Practice was reprinted here in 1781. Gregory's Lectures on the Practice were issued in 1773, and Ranby on Gunshot Wounds in 1776; and, at a later period, Benjamin Bell's System of Surgery, edited by Dr. Waters, with notes by Dr. John Jones, which went to a third edition. With the exceptions mentioned these books all issued from the press of Robert Bell, an enterprising publisher, to whom our citizens were under obligations for many valuable reprints. In the art of puffing he seems to have been in no degree behind some of his modern brethren. His advertisements of the work printed in 1775 terminates as follows: "The American physicians who wish to arrive at the top of their profession, are informed that the great Professor Cullen's Lectures on the Materia Medica, containing the *very cream* of Physic, are now selling by said Bell, in Third Street. Price five dollars."*

* Penna. Gazette, No. 24, Nov. 22, 1775.

From the above it will be seen that Bartram first led the way in selecting works on medicine for republication in America, adding to them such notes and additions as would adapt them to the wants of this country. The example was afterwards followed by Rush, in 1781, with Cullen's Practice, and Waters in 1783. At a later period Rush also introduced to his countrymen the works of Sydenham, Cleghorn, Pringle, and Hillary.

THE FOUNDATION OF HOSPITALS.

The foundation of hospitals among us produced the most important effects on the character of the medical profession, and forms a great era in our progress. The Pennsylvania Hospital, the first of these institutions established in the country, was erected principally by the contributions of the benevolent citizens of Philadelphia, though aided by a grant of £2000 from the Colonial Assembly, and received its charter in 1751. Its establishment, as has been already stated, was owing to the suggestion of a physician, Dr. Thomas Bond. Up to the period of its foundation, no college of medicine existed on the continent, and the hospital, under the care of some of the first medical men of the period, early attracted the attention of both physicians and students, and very materially contributed to the advancement and distinguished position attained by the medical school which was soon afterwards begun. In 1762 Dr. John Fothergill presented to the hospital, through William Logan, lately returned from London, a book entitled "An Experimental History of the Materia Medica," by William Lewis, F.R.S. It was given "for the benefit of the young students in physic, who may attend under the direction of the physician," and the gift seems to have led to the idea of connecting with the institution a medical library. Shortly afterwards Fothergill made another donation to the hospital of a series of anatomical drawings, framed and glazed, three cases of anatomical casts, and one case containing a skeleton and fœtus. This present, it must be remembered, was made at a time when the opportunities of acquiring anatomical knowledge were few, and when dissections were but little pursued in Philadelphia. "In want of real subjects," says Dr. Fothergill in his letter accompanying them, "these will have their use, and I have recommended to Dr. Shippen to give a course of anatomical lectures to such as may attend; he is very well qualified for the subject, and will soon be followed by an able assistant,

Dr. Morgan, both of whom, I apprehend, will not only be useful to the Province in their employments, but if suitably countenanced by the Legislature will be able to erect a school of physic amongst you that may draw many students from various parts of America and the West Indies, and at least furnish them with a better idea of the rudiments of their profession than they have at present the means of acquiring on your side of the water." This recommendation of Dr. Fothergill received the sanction of the managers, and was faithfully carried out by Dr. Shippen, and formed the first regular course upon anatomy and midwifery ever given here. They were attended by ten students, and were repeated in 1763 and 1764.

Although a number of students were attracted to the institution soon after its foundation by the reputation of its medical officers, and the advantages it afforded for the observation of disease, it was not until the year 1763 that a fee was demanded for this privilege. In that year we find the adoption of the following minute by the Board of Managers: "It being remarked that a number of students in physic do frequently attend the wards at the time of the physician visiting the patients, with a view to improve themselves in experience, it is the unanimous opinion of the Board that such of them, at least, who are not apprentices to the physician of the house should pay a proper gratuity for the benefit of the hospital for their privilege; the consideration of stipulating the sum is referred to the next board, after consulting with the physician." At the ensuing meeting of the managers the following communication was received from the physicians :—

PHILADELPHIA, May 31st, 1763.

Upon considering the minute of the Managers of the Pennsylvania Hospital made the 10th of 5th Mo. 1763, relative to those students who attend the wards of said hospital, it is our opinion that each stndent who is not an apprentice to one of the physicians attending the house shall pay six pistoles as a gratuity for that privilege. That the managers and doctors in attendance for the time being shall be the jndges who are proper to be admitted or refused. And further, as the custom of most of the hospitals in Great Britain has given such gratuities to the physicians and surgeons attending them, we think it properly belongs to us to appropriate the money arising from thence, and propose to apply it to the founding a medical library in the said hospital, which we judge will tend greatly to the advantage of the pupils, and the honor of the institution.

Signed   THOMAS BOND,
          THOMAS CADWALADER,
          PHINEAS BOND,
          CADWALADER EVANS.

(128)

"After consideration whereof the Board agrees to the proposal in respect to the terms upon which students in physic are to be admitted to attend the wards; the gratuity for which to be paid to the treasurer. And in regard to the proposal for a medical library, that such books as are purchased should be approved of by the managers, as likewise the manner in which they are to be lent out."

CLINICAL LECTURES.

The earliest attempt at formal clinical teaching ever made in the country was begun in this hospital by Dr. Thomas Bond, in 1766, whose address introductory to this course, setting forth the utility of clinical lectures, we have copied from the archives of the institution, and here present upon account of its literary and historic interest, as well as its intrinsic value.

DR. THOMAS BOND'S INTRODUCTORY LECTURE TO A COURSE OF CLINICAL OBSERVATIONS IN THE PENNSYLVANIA HOSPITAL, DELIVERED THERE THE THIRD OF DECEMBER, 1766.

"When I consider the unskilful hands the Practice of Physic and surgery has of necessity been committed to in many parts of America, it gives me pleasure to behold so

17   (129)

many worthy young men training up in these professions, which, from the nature of their objects, are the most interesting to the community; and yet a greater pleasure in foreseeing that the unparalleled public spirit of the good people of this Province will shortly make Philadelphia the Athens of America, and render the sons of Pennsylvania reputable amongst the most celebrated Europeans in all the liberal arts and sciences. This I am at present certain of, that the institutions of literature and charity already founded, and the School of Physic lately opened in this city, afford sufficient foundation for the students of physic to acquire all the knowledge necessary for their practising every branch of their professions reputably and judiciously.

"The great expense in going from America to England, and thence from country to country, and college to college, in quest of medical qualifications, is often a bar to the cultivation of the brightest geniuses amongst us, who might otherwise be morning stars in their professions, and most useful members of society. Besides, every climate produces diseases peculiar to itself, which require experience to understand and cure; and even the diseases of the several seasons in the same country are found to differ so much, some years, from what they were in others, that Sydenham, the most sagacious physician that ever lived, acknowledges that he was often difficulted and much mistaken in the treatment of epidemics for some time after their appearance.

"No country, then, can be so proper for the instruction of youth in the knowledge of physic as that in which it is to be practised; where the precepts of never failing experience are handed down from father to son, from tutor to pupil.

"That this is no speculative opinion, but real matter of fact, may be proven from the savages of America, who, without the assistance of literature, have been found possessed of skill in the cure of diseases incident to their climate superior to the regular bred and most learned physicians, and that from their discoveries the present practice of physic has been enriched with some of the most valuable medicines now in use.

"Therefore, from principles of patriotism and humanity, the physic school here should meet all the protection and encouragement the friends of their country, and well wishers of mankind, can possibly give it. Though it is yet in its infancy, from the judicious treatment of its guardians, it is already become a forward child, and has the promising appearance of soon arising to a vigorous and healthy maturity.

The professors in it at present are few, but their departments include the most essential parts of education; another,* whose distinguished abilities will do honour to his country and the institution, is expected to join them in the spring; and I think he has little faith who can doubt that so good an undertaking will ever fail of additional strength and providential blessing. And I am certain nothing would give me so much pleasure as to have it in my power to contribute the least mite towards its perfect establishment. The Professor of Anatomy and Physiology† is well qualified for the task; his dissections are accurate and elegant, and his lectures learned, judicious, and clear.

"The Professor of the Theory and Practice of Physic‡ has had the best opportunities of improvement, joined to genius and application, and cannot fail of giving necessary and instructive lessons to pupils. The field this gentleman undertakes is very extensive, and has many difficulties which may mislead the footsteps of an uncautioned traveller; therefore, lectures, in which the different parts of the Theory and Practice of Physic are judiciously classed and systematically explained, will prevent many perplexities the student would

[* Dr. Kuhn.]   [† Dr. Shippen.]   [‡ Dr. Morgan.]

otherwise be embarrassed with, will unfold the doors of knowledge, and be of great use in directing and abridging his future
studies; yet there is something further wanting; he must
join examples with study before he can be sufficiently qualified to prescribe for the sick; for language and books alone
can never give him adequate ideas of diseases and the best
method of treating them. For which reasons infirmaries are
justly reputed the grand theatres of medical knowledge.
There, the clinical professor comes into the aid of speculation,
and demonstrates the truth of theory by facts; he meets his
pupils at stated times in the hospital, and when a case pre-
sents, adapted to his purpose, he asks all those questions
which lead to a certain knowledge of the disease and parts
affected; this he does in the most exact and particular manner, to convince the students how many, and what minute
circumstances are often necessary to form a judgment of
the curative indications on which the safety and life of the
patient depend; from all which circumstances, and the present
symptoms, he pronounces what the disease is, whether it is
curable or incurable, in what manner it ought to be treated,
and gives his reasons from authority or experience for all he
says on the occasion; and if the disease baffles the power of

art, and the patient falls a sacrifice to it, he then brings his knowledge to the test, and fixes honour or discredit on his reputation by exposing all the morbid parts to view, and demonstrates by what means it produces death, and if, perchance, he finds something unexpected, which betrays an error in judgment, he, like a great and good man, immediately acknowledges the mistake, and for the benefit of survivors points out other methods by which it might have been more happily treated. The latter part of this field of tuition is the surest method of obtaining just ideas of diseases. The great Boerhaave was so attentive to it, that he was not only present at the opening of human bodies, but frequently attended the slaughter-houses in Leyden, to examine the carcasses of beasts; and being asked by a learned friend by what means he acquired such uncommon certainty in the diagnostics and prognostics of disease answered, 'By examining dead bodies, studying Sydenham's Observations, and Bonetus's Sepulchretum Anatomicum,' both of which he had read ten times, and each time with greater pleasure and improvement.

"But to give you more familiar instances of the utility of this practice, let me remind several of you, who were present last fall at the opening of two bodies, one of which died of asthmatic complaints, the other of a frenzy, succeeded by a palsy, and ask you if anything short of ocular demonstration could have given you just ideas of the causes of the patient's death; in one we saw a dropsy in the left side of the thorax, and a curious polypus with its growing fimbriæ of fourteen inches in length (now in the hospital), extending from the ventricle of the heart, far beyond the bifurcation of the pulmonary artery; in the other we found the brain partly suppurated, and the ventricle on the opposite side to that affected with the paralysis distended by a large quantity of limpid serum, and you must remember that the state of all the morbid parts was predicted before they were exposed to view, which may have a further advantage, by arousing in you an industrious pursuit after the most hidden causes of all the affections of the human body, and convince you what injury they do the living who oppose a decent, painless, and well-timed examination of the dead.

"Thus, all the professors in the best European colleges go hand in hand, and co-operate with each other by regular chains of reasoning and occasional demonstrations, to the satisfaction and improvement of the students.

"But more is required of us in this late settled world,

where new diseases often occur, and others, common to many parts of Europe, visit us too frequently, which it behooves the guardians of health to be very watchful of, that they may know them well, and by a hearty union and brotherly communication of observations, investigate their causes, and check their progress. The task is arduous, but it is a debt we owe to our friends and our country. The atmosphere which surrounds us is fine, and the air we breathe free, pure, and naturally healthy, and I am fully persuaded we shall find, on strict inquiry, when it becomes otherwise, it is mostly from contagion imported, or neglected sources of putrefaction amongst ourselves, and, therefore, whenever we are able to demonstrate the causes, they may be removed, and the effects prevented.

"Our fathers, after insuring to us the full enjoyment of the inestimable blessings of religious and civil liberty, have settled us in a country that affords all the real comforts of life, and given us the prospect of becoming one day a great and happy people; and I know only one objection to a prudent man's giving North America the preference to any other part of the British Dominions for the place of his residence, which is, that the climate is sometimes productive of severe epidemic diseases in the summer and fall. The country is otherwise free from those tedious and dangerous fevers which frequently infest most parts of Europe. The last wet summer, and short space of hot, dry weather in autumn, caused so many intermittents from the southern suburbs of this city all the way to Georgia, that I may venture to assert two-thirds of the inhabitants were not able to do the least business for many weeks, and some families, and even townships, were so distressed that they had not well persons sufficient to attend the sick, during which time this city was unusually healthy. How respectable, then, would be the character of those men, who should wipe this stain out of the American escutcheon, and rescue their country from such frequent calamities.

"Sufficient encouragement to make the attempt is found both in History, the Books of Physic, and our experience. Several instances are recorded of places that were so sickly, as to be uninhabitable, until Princes have ordered their Physicians to search into the causes of their unhealthiness; and having discovered and removed them, made thereby valuable additions to their Kingdoms. Was not our antient and great master, Hippocrates, so knowing in the cause of pestilential contagion, as to foresee an approaching plague, and send his pupils into the cities to take care of the sick? and has not he, and Sydenham, the English Hippocrates, done infinite service to the healing Art, and gained immortal Honours to themselves by their essays on epidemics, in which they not only accurately describe the diseases of their respective countries, but show the depraved constitution of the air which produced each of them? Our own experience also affords much encouragement: when I first came into this city, the Dock was a common sewer of filth, and was such a nuisance to the inhabitants about it that every Fall they were obliged to use more pounds of bark than they have ounces since it has been raised and levelled. Another striking instance of the advantage of cleanliness for the preservation of health, affords me an opportunity of paying a tribute, justly due, to the wisdom of the Legislature of this Province, in framing the salutary laws for paving and regulating the streets of this city, and to the indefatigable industry and skill of the commissioners in executing them; whereby, they have contributed so much to the healthiness of the inhabitants that I am confident the whole expense will be repaired in ten years by the lessening of the physic bills alone. A farm within a few miles of this city was remarkably healthy for fifty years, whilst the tide overflowed the lowlands, near the dwelling-house; but after they were banked in by ditches so ill contrived that they did not often discharge the water that fell into them for a considerable time, and until it became putrid, and thereby rendered the place as remarkably sickly as it had been before healthy, I was told by a gentleman of veracity, that he saw the corpses of nine tenants that had been carried from it in a few years.

"The yellow fever, which I take to be exactly the same distemper as the plague of Athens, described by Thucydides, has been five different times in this city since my residence in it; the causes of three of them I was luckily able to trace, and am certain they were the same which produced a gaol fever in other places, and am of opinion the difference betwixt the appearance of these fevers arises from the climate, and the different state the bodies are in when they imbibe the contagion; if so, the same methods which are taken to prevent a gaol fever will equally prevent a yellow fever: it was in the year forty-one I first saw that horrid disease, which was then imported by a number of convicts from the Dublin gaol. The second time it prevailed it was indigenous, from evident causes, and was principally confined to one square of the city. The third time it was generated on board of crowded ships in the port, which brought their passengers in health, but soon after became very sickly. I here saw the appearance of contagion like a dim spark, which gradually increased to a blaze, and soon after bursts into a terrible flame, carrying devastation with it, and after continuing two months, was extinguished by the profuse sweats of tertian fevers; but this is not the ordinary course of the contagion, it is usually checked by the cool evenings in September, and dies on the appearance of an October frost.

"I lately visited an Irish passenger vessel, which brought the people perfectly healthy until they came in our river; I found five of them ill, and others unwell, and saw that the fomes of infection was spreading among them; I, therefore, ordered the ship to lay quarantine, to be well purified with the steams of sulphur, and with vinegar; directed the bedding and clothing of the people to be well washed and dried before any person should be permitted to land out of her; after which I advised separating the sick from the healthy. This was done by putting twelve in different rooms in one house, and fourteen in another, out of the city; the con-

veniences of the two houses were much the same; in one of
them little care was taken of the sick, who were laid upon
the same foul beds, they (contrary to orders) brought on shore
with them: the consequence was, that all the family catched
the distemper, and the landlord died. In the other, my directions were strictly observed; the sick had clean clothes and
clean bedding, were well attended, and some recovered without doing the least injury to any person that visited them;
which confirms observations I had made before, that the contagion of malignant fevers lies in the air, confined and
corrupted, by neglect of rags and other filth about the help-
less sick, and not from their bodies.

"As these heads shall be the subject of a future lecture, I
shall at present only mention to you further, a few of those
methods which have preserved individuals from prevailing
diseases.

"The inhabitants of Hispaniola have found the wearing of
flannel shirts to be a preservative against intermitting fevers
in that sickly island; and as that disease is known to arise
principally from inhaling a great quantity of the humidity of
the air, I make no doubt it would also be of use in preventing
them in our low, moist, level countries.

" We know that the bark of sassafras contains many excellent medicinal virtues: my worthy friend, Mr. Peter Franklin, told me, that he being, in the Fall of the year, in the river Nanticoke, in Maryland, and on seeing the people on shore much afflicted with intermittent fevers, advised the mariners of the ship to drink freely, by way of prevention, of that aromatic and antiseptic medicine, but could not prevail on more than half the company to do it, and that he and all others who took it enjoyed perfect health, whilst not a single person of the rest escaped a severe attack of the epidemic disease: I have known other similar instances, which it is needless to mention, since this is remarkably pertinent.

" But I have many reasons to expect that a more agreeable and equally certain preventative against our autumnal fevers will be found in sulphurous chalybeate waters, which may be readily procured in most parts of America, especially where those diseases are most prevalent: a spring of this kind at Gloucester, within a few miles of this place, has been much used of late, has been so very serviceable to invalids, it has the appearance of being a valuable conveniency to the city. Persons under various diseases took lodgings in the village last season, for the advantage of drinking the waters at the

fountain head, and though the Fall was more sickly than has
been known in the memory of man, not any one of the inhabitants near the Spaw, who drank freely, had a touch of
the prevailing disease, whilst the major part of those that
did not, had more the appearance of ghosts than living crea-
tures.   There were two houses, the habitations of father and
son, within twenty feet of each other; the family of the father
had suffered greatly from intermitting fevers the preceding
Fall, and some of them continued invalids till the middle of
Summer, when they were prevailed on to take the waters,
after which they daily recovered health, bloom, and vigour, and
passed the sickly season without a complaint; whilst scarcely
a person in that of the son, who did not take them, escaped
a severe illness.   It is well known from experience, that min-
eral waters are not only the most palatable, but the most salutary parts of the materia medica; and that the effect of those
which are pure and properly impregnated with the chalybeate
principles, strengthen digestion, brace and counteract the
summer's sun, dilute a thick, putrid bile (the instrument of
mischief in all hot climates), and immediately wash away
putrefaction through the emunctories of the bowels, skin, or
kidneys, and therefore appear to be natural preservatives

against the effect of a hot, moist, and putrid atmosphere. Whether these waters will answer my sanguine expectations or not must be left to the decision of time; if they should be found wanting, that ought not discourage our further pursuit; for since Providence has furnished every country with defences for the human body against the inclemencies of heat and cold, why should we question whether Infinite Wisdom and Goodness has made equal provision against all the natural injuries of our constitution? Experience and reason encourage us to believe it has, and that the means might be discovered by diligent investigation were our researches equal to the task. The above instances are, therefore, related to convince you that the prevention of some of the epidemic diseases of America is not only a laudable and rational pursuit, but is more within the limits of human precaution than has generally been imagined; and to excite your particular attention to the improvement of this humane and interesting part of your profession, in which, and all other useful undertakings, I most sincerely wish you success.

"I am now to inform you, gentlemen, that the Managers and Physicians of the Pennsylvania Hospital, on seeing the great number of you attending the School of Physic in this city, are

of opinion, this excellent institution likewise affords a favourable opportunity of further improvement to you in the practical part of your profession; and being desirous it should
answer all the good purposes intended by the generous con-
tributors to it, have allotted me the task of giving a course of
clinical and meteorological observations in it, which I cheer-
fully undertake (though the season of my life points out
relaxation and retirement, rather than new incumbrances), in
hopes that remarks on the many curious cases that must daily
occur amongst an hundred and thirty sick persons collected
together at one time may be very instructive to you, I
therefore purpose to meet you at stated times here, and give
you the best information in my power of the nature and
treatment of chronical diseases, and of the proper management of ulcers, wounds, and fractures. I shall show you all
the operations of surgery, and endeavour, from the experience
of thirty years, to introduce you to a familiar acquaintance
with the acute diseases of your country; in order to which,
I shall put up a complete meteorological apparatus, and
endeavour to inform you of all the known properties of the
atmosphere which surrounds us, and the effects its frequent
variations produce on animal bodies; and confirm the doctrine,

by an exact register of the weather, and of the prevailing diseases, both here and in the neighbouring provinces; to which I shall add all the interesting observations which may occur in private practice, and sincerely wish it may be in my power to do them to your satisfaction.

"I have, likewise, the pleasure to inform you that Dr. Smith* has promised to go through a course of experimental philosophy in the college, for your instruction on Pneumatics, Hydraulics, and Mechanics, which will be of the greatest advantage to a ready comprehension of the meteorological lectures, and other parts of your Medicinal studies, and lay you under the highest obligations to that learned professor."

Connected with Dr. Bond in this work of instruction must also be mentioned the venerated names of his associates, Cadwalader, Evans, Redman, and Morgan, all of whom entered heartily into the plan of clinical teaching.

The fee arising from the privilege granted to students of attending the hospital practice was at this period increased from six pistoles to a guinea, to be applied to the promotion of the medical library, which now comprises a large propor-

[* The Provost.]

tion of the most valuable ancient and modern writers on the science of medicine, together with many rare works on botany, and the different branches of natural history, and stands a noble monument to the zeal, liberality, and love of science of her medical men.

To the Pennsylvania Hospital, Philadelphia is much indebted for the reputation which she has enjoyed for medical and surgical teaching. From her halls in later years emanated the valuable practical lessons of Kuhn, and the eloquent instructions of Rush, while Shippen, Jones, Physick, and Wistar, names among the brightest and most revered in early American surgery, found there a field for practice, and opportunities to disseminate widely sound surgical principles. Oldest as she is in our practical schools, she can boast of having aided largely in the progress and improvement of medical science in America. In her well-regulated wards, students have always learned, both by precept and practice, that the minor duties of a surgeon, and the common accidents to which the human family are exposed, deserve most of their attention, and in her theatre doubtful operative schemes, rash experiments, or mere bloody exploits have ever been condemned, while, at the same time, her records show, that bold achievements, when founded on well-established principles, have frequently been witnessed, and, at times, originated there.

FOUNDATION OF MEDICAL SCHOOLS.

Philadelphia has the high honour of giving birth to the first medical school in the New World, and the scheme, we feel proud to say, originated with, and was carried into execution by, one of our own town's-people. To the Academy of Philadelphia, afterwards known as the College, and now bearing the title of the University of Pennsylvania, belongs the credit of having opened this new path of liberal study. As the pioneer in the great field of medical instruction, she could not be passed over without some notice by the medical annalist, and when, in addition to this, we consider the public spirit, learning, and ability of her founder and professors, the distinguished position which she has so long held, the honour her teachers have brought upon the American name, the influence which she has exerted upon the profession, and the important benefits which the whole country have derived from her, some details of the rise of this great school may not here be misplaced.

Previous to the formation of the medical department of the

College of Philadelphia, the city, as we have shown, could boast of many well educated and, indeed, eminent practitioners, and as we are told by Morgan,* was already the resort "of a great number of pupils from the neighbouring parts, to learn the arts of Physic and Surgery." The best medical education, however, in America was then but lame and insufficient. Under some skilful surgeon or physician it was customary for the student to enter as an apprentice, in order to learn their mode of practice, and this, together with casual conversations and intercourse with one another and professional men, and a reciprocal communication of sentiment and observation, together with reading such authors as they could procure on the various subjects of our science, made the sum total of it. Such was the state of medical education in 1762, when Dr. Shippen arrived from Europe, who soon afterwards commenced his demonstrative lectures upon Anatomy with dissections. These lectures he repeated annually till the year 1765, when Dr. John Morgan returned.

This gentleman had matured a project for connecting a medical department with the College of Philadelphia, and had

---

* A Discourse upon the Institution of Medical Schools in America. p. 30.

secured in favour of it some influential friends of the institution, who at that time were in England. Among these were Thomas Penn, the proprietor of Pennsylvania, James Hamilton, and the Rev. Mr. Peters, former members of the Board of Trustees of the College, from all of whom he brought letters advising the establishment of medical professorships, and recommending the doctor himself to their choice as one of the proposed faculty. The following is the letter from Mr. Penn :—

GENTLEMEN :

"Dr. Morgan has laid before me a proposal for introducing new professorships into the college, for the instruction of such as shall incline to go into the study and practice of physic and surgery, as well as the several occupations attending upon these necessary and useful arts. He thinks his scheme, if patronized by the Trustees, will at present give reputation and strength to the institution, and though it may for some time occasion a small expense, yet after a little while it will gradually support itself, and even make considerable additions to the Academy funds.

"Dr. Morgan has employed his time in an assiduous search after knowledge, in all the branches necessary for the practice of his profession, and has gained such esteem and love from persons of the first rank in it, that as they very much approve of his plan, they will from time to time, as he assures us, give him their countenance and assistance in the execution of it.

"We are made acquainted with what is proposed to be taught, and how lectures may be adopted by you; and since the like systems have brought much advantage to every place where they have been received, and such learned and eminent men speak favourably of the doctor's plan, I could not but in the most kind manner recommend him to you, and desire that he may be well received, and what he has to offer to be taken with all becoming respect and expedition into your most serious consideration; and if it shall be thought necessary to go into it, and thereupon to open professorships, that he may be taken into your service.

"When you have heard him, and duly considered what he has to lay before you, you will be best able to judge in what manner you can serve the public, the institution, and the particular design now recommended to you.

"I am, gentlemen,

"Your affectionate friend,

"THOMAS PENN.

"LONDON, February 15th, 1765.
"To the Trustees of the College of Philadelphia."

The letters brought by Dr. Morgan, together with his plan "of opening medical schools under the patronage and government of the College," were laid before the Trustees on the 3d of May, 1765. The project met their unanimous approval, and Dr. Morgan was elected to the professorship of the Theory and Practice of Physic, for which he had applied.

The official announcement of this appointment was as follows :—

The Early History of Medicine in Philadelphia.

"At a meeting of the Trustees held this day, John Morgan, of this city, M.D., corresponding member of the Royal Academy of Surgery at Paris, Licentiate of the Royal College of Physicians at London, and member of the Arcadian (Belles Lettres) Society at Rome, was unanimously elected Professor of the Theory and Practice of Medicine in the College of Philadelphia. At the ensuing commencement he will deliver an address (which will be soon afterwards published), in order to show the expediency of instituting medical schools in this seminary, and containing the plan proposed for the same; in which there will be room for receiving professors duly qualified to read lectures in the other branches of medicine, who may be desirous of uniting to carry this laudable design into execution. DR. MORGAN'S PLAN has been warmly recommended to the Trustees by persons of eminence in England, and his known abilities and great industry give the utmost reason to hope it will be successful, and tend much to the public utility."

At the collegiate commencement, which took place on the 30th of May, he delivered his "Discourse upon the Institution of Medical Schools in America." This address, which

he had carefully prepared while in Paris, had been submitted
before his return to Drs. Fothergill, Hunter, and Watson, of
London, as well as to the criticism of his friend and fellow-
traveller, Samuel Powel, Esq., a wealthy and distinguished
native of our city, who aided materially in carrying out his
projects. In this able production he first states the various branches of knowledge which compose the science of
medicine, and the order in which they should be studied. He
then dwells upon the importance of preliminary education,
and urges that "young men ought to come well prepared for
the study of medicine, by having their minds enriched with
all the aids they can receive from the languages, mathematics,
and all the liberal arts." A brief general view of the state
of physic, as hitherto practised in America, and the obstacles
to its study are then given, with forcible arguments in favour
of instituting medical schools among us. The many circumstances conspiring to invite and encourage so important an
undertaking, among which he enumerates the eminent practitioners who adorn our city, the existence of the hospital
which had been established ten years before, the flourishing
state of literature, the favourable central situation of Philadelphia to the colonies, and the great growth of all the latter

20   (153)

in population, are then all alluded to, and after expatiating on the advantages which are to be expected from the proposed Institution to students of medicine, to the College, to the city of Philadelphia, and the Province, as well as to the neighbouring colonies, he addresses himself to the students present, in order to animate them in their studies, and closes with an appeal to the Trustees of the College, recommending the medical school to their patronage. "Perhaps," says Dr. Morgan in this address, "this Medical Institution, the first of its kind in America, though small in the beginning, may receive a constant increase of strength, and annually exert new vigour. It may collect a number of young persons, of more than ordinary abilities, and so improve their knowledge as to spread its reputation to distant parts. By sending these abroad duly qualified, or by exciting an emulation amongst men of parts and literature, it may give birth to other useful Institutions of a similar nature, or occasional rise, by its example, to numerous societies of different kinds, calculated to spread the light of knowledge through the whole American Continent, wherever inhabited."

.   .   .   .   .

"Oh! let it never be said in this city, or in this Province, so happy in its climate and its soil, where commerce has long flourished and plenty smiled, that science, the amiable daughter of liberty and sister of opulence, droops her languid head, or follows behind with a slow unequal pace. I pronounce with confidence this shall not be the case; but under your protection every useful kind of learning shall here fix a favourite seat, and shine forth in meridian splendour. To accomplish which may every heart and every hand be firmly united."

> Hoc opus, hoc studium parvi properemus et ampli
> Si patria volumus, si nobis vivere cari. *Hor. Epist.*

The discourse was pronounced at two sittings, on the 30th and the 31st of the month. The place selected for its delivery was the old Academy on Fourth Street near Arch. The speaker appeared in a professor's robe, and a large concourse of leading citizens, including the Governor, assembled to listen to it. Franklin's newspaper, in noticing it, says: "As it is soon to be printed, we would not wish to anticipate the judgment of the public, and shall only say, the perspicuity with which it was written and spoken drew the close attention of the audience, and particularly of the gentlemen of the Faculty of Physic."

The Early History of Medicine in Philadelphia.

On the 23d of September of the same year (1765), Dr. Shippen addressed the following communication to the Trustees :—

"To The Trustees of the College, etc.

"The institution of medical schools in this country has been a favourite object of my attention for seven years past, and it is three years since I proposed the expediency and practicability of teaching medicine in all its branches in this city, in a public oration, read at the State House, introductory to my first course of anatomy.

"I should long since have sought the patronage of the Trustees of this College, but waited to be joined by Dr. Morgan, to whom I first communicated my plan in England, and who promised to unite with me in every scheme we might think necessary for the execution of so important a point. I am pleased, however, to hear that you gentlemen, on being applied to by Dr. Morgan, have taken the plan under your protection, and have appointed that gentleman Professor of Medicine.

"A professorship of Anatomy and Surgery will be gratefully accepted by, gentlemen, your most obedient and humble servant,

"WILLIAM SHIPPEN, Jr.

"Philadelphia, 17th September, 1765."

Which letter being read, the Trustees, by an unanimous vote, appointed Dr. Shippen Professor of Anatomy and Surgery.

(156)

The courses in the school thus organized were soon afterwards regularly commenced, and the advertisement announ-
cing this first collegiate course of medical instruction in
America is annexed, as worthy of preservation, and as showing the scheme of instruction proposed.

"As the necessity of cultivating medical knowledge in
America is allowed by all, it is with pleasure we inform the
public that courses of lectures on two of the most important
branches of that useful science, viz., Anatomy and Materia
Medica, will be delivered this winter in Philadelphia. We
have great reason, therefore, to hope that gentlemen of the
faculty will encourage the design by recommending it to their
pupils, that pupils themselves will be glad of such an opportunity of improvement; and that the public will think it an
object worthy their attention and patronage. In order to
render these courses the more extensively useful, we intend to
introduce into them as much of the theory and practice of
physic, of pharmacy, chemistry, and surgery as can conve-
niently be admitted. From all this, together with an attendance on the practice of the physicians and surgeons of the
Pennsylvania Hospital, the students will be able to prosecute
their studies with such advantages as will qualify them to

practice hereafter with more satisfaction to themselves, and benefit to the community. The particular advertisements inserted below, specify the time when these lectures are to commence, and contain the various subjects to be treated of in each course, and the terms on which the pupils are to be admitted.

"WILLIAM SHIPPEN, Jun., M.D.,
"Professor of Anatomy and Surgery in the College of Philadelphia.

"JOHN MORGAN, M.D., F.R.S.,
"And Professor of Medicine in the College of Philadelphia.

"Dr. Shippen's course of anatomical lectures will begin on Thursday, the 14th of November, 1765; it will consist of about sixty lectures, in which the situation, figure, and structure of all parts of the human body will be demonstrated on the fresh subject, their respective uses explained, and their diseases, with the indications and methods of cure, briefly treated of; all the necessary operations on Surgery will be performed, a course of bandages given; and the whole concluded with a few plain and general directions in the practice of midwifery. Each person to pay six pistoles.

"Those who incline to attend the Pennsylvania Hospital,

and have the benefit of the curious anatomical plates and casts there, to pay six pistoles to that useful charity.

"A course of lectures on the Materia Medica by John Morgan, M.D., etc. Price four pistoles.

"This course will commence on Monday, the 18th day of November, and be given three times a week at the College, at three o'clock in the afternoon, till finished, which will last between three and four months.

"To render these lectures as instructive as possible to students of physic, the Doctor proposes, in the course of them, to give some useful observations on medicine in general, and the proper manner of conducting the study of physic. The authors to be read in the Materia Medica will be pointed out. The various substances made use of in medicine will be reduced under classes suited to the principal indications in the cure of diseases. Similar virtues in different plants, and their comparative powers will be treated of, and an inquiry made into the different methods which have been used in discovering the qualities of medicines; the virtues of the most efficacious will be particularly insisted upon; the manner of preparing and combining them will be shown by some instructive lessons upon pharmaceutic chemistry. This will open to students a general idea of both chemistry and pharmacy. To prepare them more effectually for understanding the art of prescribing with elegance and propriety, if time allows, it is proposed to include in this course some critical lectures upon the chief preparations contained in the Dispensatories of the Royal Colleges of Physicians at London and Edinburg. The whole will be illustrated with many useful practical observations on Diseases, Diet, and Medicines.

"No person will be admitted without a ticket for the whole course. Those who propose to attend this course are desired to apply to the Doctor at least a week before the lectures begin. A dollar will be required of each student to matriculate, which will be applied to purchase books for a medical library in the College, for the benefit of the medical students."

Much discrepancy of opinion has existed as to whom is due the honour of founding our medical school, and usually, though erroneously, it has been awarded to Dr. Shippen. It is true that introductory to his first course of lectures on Anatomy, delivered in the State House in 1762, he spoke of the institution of a medical school in Philadelphia, to which he then declared that a course of anatomy should be introductory. The use of such an institution, and the practicability and propriety of it at that time, were expressed in these words:—

"All these (alluding to the disadvantages that then attended the study of physic, etc.) may, and I hope will soon be remedied by a medical school in America; and what place in America so fit for such a school as Philadelphia, that bids so fair by its rapid growth to be soon the metropolis of all the continent? Such a school is properly begun by an anatomical class, and for our encouragement, let us remember that the famous school of physic at Edinburg, which is now the first in Europe, has not had a beginning fifty years, and was begun by the anatomical lectures of Dr. Monro, who is still living."

But on turning to the Discourse of Dr. Morgan, delivered before the Trustees in 1765, we find he awards full credit to Shippen for these efforts. He says: "It is with the highest satisfaction I am informed from Dr. Shippen, junior, that in an address to the public as introductory to his first anatomical course, he proposed some hints of a plan for giving medical lectures amongst us. But I do not learn that he recommended at all a collegiate undertaking of this kind. What led me to it was the obvious utility that would attend

21   (161)

it, and the desire I had of presenting, as a tribute of gratitude to my Alma Mater, a full and enlarged plan for the institution of medicine, in all its branches, in this seminary, where I had part of my education, being amongst the first sons who shared in its public honours.* I was further induced to it from a consideration that private schemes of propagating knowledge are unstable in their nature, and the cultivation of useful learning can only be effectually promoted under those who are patrons of science, and under the authority and direction of men incorporated for the improvement of literature."

He then goes on to state, that should the Trustees think proper to found a professorship of anatomy, "Dr. Shippen having been concerned already in teaching that branch of medical science, is a circumstance favourable to our wishes. Few here can be ignorant of the great opportunities he has had abroad of qualifying himself in anatomy, and that he has already given three courses thereof in this city, and designs to enter upon a fourth course next winter."

From these statements of the rivals for the honour, there

---

* He was a member of the first class that graduated in the Collegiate Department of the College, 1757.

can be no doubt about their respective claims. Dr. Shippen introduced and successfully taught a single branch of medicine in 1762, but took no steps for the establishment of a medical institution upon a permanent basis. Dr. Morgan arrived at home in April, 1765, and in the following month proposed to the Trustees of the College HIS PLAN for transplanting medical science into their seminary, and boldly urged upon them a full and enlarged scheme for the institution of medicine in all its branches.

It is thus evident that Dr. Morgan must be considered as the founder of the first medical school in America. His aim was to establish one similar to that of Edinburg, and he engaged in it, as we have seen, with all the industry and ability necessary for the accomplishment of so important an object. Dr. Shippen was a valuable and most useful adjunct, and much credit is due to him for his vigorous prosecution of anatomical pursuits, his demonstrations on midwifery, and the zeal and ability which he brought to aid in the difficult undertaking.*

* Dr. John Redman Coxe has kindly placed in my hands some letters from the venerable Dr. Redman to Dr. Morgan, then in Europe, from one of which I extract the following :—

(163)

The Early History of Medicine in Philadelphia.

In May, 1767, two years after the school had been in operation, rules were adopted in relation to the medical honours to be conferred, which were announced as follows:—*

"PHILADELPHIA, March 13, 1764.

"This leads me to make some reflections on your intended plan, which I still approve, as far as useful and practicable. As to that part of it which relates to instruction by lectures, etc., it is highly praiseworthy, and should by all means be attempted, and you'll thereby deserve much honour for so important a service to this part of the world, and will no doubt obtain it, and will have just reason to think yourself upon every good principle justly entitled thereto, for your pains and expense, and enjoy a conscious pleasure of having endeavoured to serve mankind in their best temporal interests."

In a letter written from London, Nov. 10th, 1764, to his former preceptor, Dr. Cullen, Morgan says, "My scheme for instituting lectures, you will hereafter know more of. It is not prudent to broach designs prematurely, and mine are not yet fully ripe for execution."

In a publication made by Dr. Morgan, in 1781—sixteen years after its first establishment—he asserts himself to have been "the original founder," and in the same year, Dr. Rush, than whom no one had a better opportunity of knowing, speaks of Morgan as "a man to whom America is indebted for the foundation of her first Medical School."

---

* Five of the six prominent physicians of Philadelphia were Trustees of the College at this period, and they united with the Provost, Dr. William Smith, and the two newly made medical professors, in digesting this code.

The Early History of Medicine in Philadelphia.

"COLLEGE OF PHILADELPHIA, July 29th, 1767.

"At a meeting of the Trustees, held the twelfth day of May last, it being moved to the Board, that conferring the usual Degrees in Physic on deserving students would contribute greatly to encourage the medical school, lately instituted in this seminary, promote emulation among the students, and tend to put the Practice of Physic on a more respectable footing in America; the motion was unanimously agreed to, and the following course of studies and qualifications, after mature deliberation, was fixed on and enacted as requisite to entitle physical students to their different degrees, viz:—

"FOR A BACHELOR'S DEGREE IN PHYSIC.

"1. It is required that such students as have not taken a degree in any College shall, before admission to a Degree in Physic, satisfy the Trustees and Professors of the College, concerning their knowledge in the Latin Tongue, and in such branches of Mathematics, Natural and Experimental Philosophy, as shall be judged requisite to a Medical Education.

"2. Each student shall attend at least one course of lectures in Anatomy, Materia Medica, Chemistry, the Theory and Practice of Physic, and one course of Clynical lectures, and shall attend the practice of the Pennsylvania Hospital for one year; and may then be admitted to a public examination for a Bachelor's Degree in Physic; provided, that in a previous private examination by the Medical Trustees and Professors, and such other Trustees or Professors as choose to attend, such students shall be judged fit to undergo a public examination without attending any more courses in the Medical School.

"3. It is further required that each student, previous to the Bachelor's Degree, shall have served a sufficient apprenticeship to some respectable practitioner in Physic, and to make it appear that he has a general knowledge in Pharmacy.

"QUALIFICATIONS FOR A DOCTOR'S DEGREE IN PHYSIC.

"It is required for this Degree that at least three years shall have intervened from the time of taking the Bachelor's Degree, and that the candidate be full twenty-four years of age, and that he shall write and defend a Thesis publickly in College, unless he should be beyond seas, or so remote on the Continent of America as not to be able to attend without manifest inconvenience; in which case, on sending a written

Thesis, such as shall be approved by the College, the candidate may receive his Doctor's Degree, but his Thesis shall be printed and published at his own expense.

"FEES TO PROFESSORS.

"No Professor to take more than six Pistoles for a single course, in any of the above branches, and after two courses, any student may attend as many more as he pleases gratis."

"This scheme of a Medical Education is proposed to be on as extensive and liberal a plan, as in the most respectable European seminaries, and the utmost provision is made for rendering a Degree, a real mark of honour, the reward only of distinguished Learning and Abilities. As it is calculated to promote the benefit of mankind, by the improvement of the beneficent art of Healing, and to afford an opportunity to students of acquiring a regular medical education in America, it is hoped it will meet with public encouragement, more especially as the central situation of this city, the established character of the Medical Professors, the advantages of the College, and the public hospital all conspire to promise success to the design.

"For the further advantage of medical students, a course

of lectures will be given by the Professor of Natural and Experimental Philosophy each winter in the College, where there is an elegant and compleat apparatus provided for that purpose; and where medical students may have an opportunity of compleating themselves in languages, or any parts of the Mathematics at their leisure hours.

"Agreeably to the foregoing regulations the public is now informed, that on the first Monday in November next, the following courses of lectures will be begun by the respective Professors, viz:—

"A compleat course of lectures on Anatomy, to which will be added all the operations in surgery, and the mode of applying all the necessary bandages, etc. A course of lectures on the Theory and Practice of Medicine, which will be preceded by a general explanation of the theory of Chemistry, accompanied with some necessary operations, to render a knowledge of this science easy and familiar to the inquisitive student.

"A course of clynical lectures, to be delivered in the Pennsylvania Hospital, wherein the treatment of both acute and chronic diseases will be exemplified in the cases of a great number of patients.

"Each course of lectures will be finished by the beginning

of May, in time for those who intend to offer as candidates for a Degree in Physic, to prepare themselves for the examination before the commencement of the ensuing year.

"Such gentlemen as incline to attend the above courses are desired to apply some days before the Lectures begin, to furnish themselves with the necessary tickets of admission."

Bachelor's degrees were found by experience not to answer the end proposed by them. Of the many who had taken the degree of Bachelor, very few returned to apply for the highest honour in medicine, and in consequence the system of granting Bachelorships was abolished, and from November 17th, 1789, the degree of M.D. only was recognized.

Dr. Morgan had much doubt about the propriety of conferring a Bachelor's degree originally, and consulted Cullen and other of his old teachers in regard to it. Dr. Rush, who was then in Edinburg, was also consulted upon the matter, and in a letter to Dr. Morgan, dated April 27, 1768, says :—

"I have read the laws you have established with regard to the conferring of Degrees in Physic, and have shown them to several gentlemen in this place, who, upon the whole, approve of them. Some have thought that conferring Bachelor's degrees in Physic would tend to depreciate their value, as very few young men would ever have leisure enough after they begin to practice to return a second time to the College in order to write a Thesis, or go through the other necessary forms previous to their being admitted 'Doctors of Physic.' On this account they have proposed that no one shall be admitted to the physical honours of the College, until they have studied there two or three years, and afterwards published a Thesis. But you who are on the spot can judge best of the propriety of such a regulation."

The requirements for graduation were also at this time somewhat changed. The principal of these changes were, that the candidate should have reached the age of twenty-one years; that he should have studied medicine three years, two of which were to be in connexion with the University; that he should undergo a private examination before the faculty; and that he should hand in a Thesis to be defended at the public commencement, and to be printed at his own expense. These requirements continued to be demanded till 1807, when they were still further changed for those now in force.

"In January 26th, 1768, Dr. Adam Kuhn having made application to be appointed Professor of Botany and Materia

Medica in this College, declaring that he would do the utmost in his power to merit that honour, and the Trustees, having ample assurance of his abilities to fill that Professorship, for which he is likewise particularly recommended by the Medical Trustees and Professors belonging to the College, do therefore unanimously choose and appoint him, the said Dr. Adam Kuhn, to be Professor of Botany and Materia Medica in this College agreeably to his request."

In the following May, Dr. Thomas Bond, who was already engaged in clinical teaching at the Pennsylvania Hospital, was elected to the chair of Clinical Medicine, a position which was held by him until his death in 1784, after which it was united with that of the Institutes of Medicine.

The commencement of the College was held in June, 1768,* when medical honours were conferred for the first time in America. The following account of the exercises on that

* This was three years after the foundation of the College. The first degrees conferred by King's College, in New York, were in 1769, one year after the opening. They were both degrees of Bachelor of Medicine. The first degree of Doctor of Medicine ever conferred in this country was an honorary one in 1723, granted by Yale College to Daniel Turner, of London, author of "The Art of Surgery," who had sent out to that institution a collection of valuable books in physick and surgery.

occasion, which I copy from the newspapers of the day, will perhaps not be uninteresting:—

"COLLEGE OF PHILADELPHIA, June 21, 1768.

"This day, which may be considered as having given Birth to Medical Honours in America, the following Gentlemen were admitted to the degree of Bachelor of Physick, viz: Messieurs John Archer, of Newcastle County; Benjamin Cowell, of Bucks County; Samuel Duffield, of Philadelphia; Jonathan Elmer, of West Jersey; Humphrey Fullerton, of Lancaster County; David Jackson, of Chester County; John Lawrence, of East Jersey; Jonathan Potts, of Pottsgrove; James Tilton, of Kent County; and Nicholas Way, of Wilmington, New Castle County.

"Agreeably to the rules of the College, these Gentlemen, previous to their admission to a Degree, had diligently attended the lectures of the several Professors in Anatomy, the Materia Medica, Chemistry, Theory and Practice of Physick, and the Clinical lectures in the Pennsylvania Hospital; in which (as well as in the *Languages* and the necessary branches of *Natural Philosophy*) they gave the most satisfactory proofs of their proficiency, both in their private and publick examinations.

"The Provost, after opening the commencement with prayers, introduced the business of the day with a short Latin Oration. Then followed,

"1. A Latin oration delivered with great propriety by Mr. Lawrence, 'De Honoribus qui omni Œvo in veros Medicinæ Cultores collati fuerunt.'

"2. A dispute 'Whether the Retina or Tunica Choroides be the immediate seat of vision?' The argument for the Retina was ingeniously maintained by Mr. Cowell. The opposite side of the question was supported with great acuteness by Mr. Fullerton, who contended that the Retina is incapable of the office ascribed to it, on account of its being easily permeable by the rays of light, and that the choroid coat, by its being opake, is the proper part for stopping the rays, and receiving the picture of the object.

"3. Questio 'Num datur Fluidum Nervosum?' This question was discussed with great learning; the affirmative by Mr. Duffield, and the negative by Mr. Way.

"4. Mr. Tilton delivered an essay 'On Respiration, and the manner in which it is performed,' which did great credit to his abilities.

"5. The Degrees were conferred by the Provost, after

receiving the mandate of the Trustees from his Honour the Governor, as their President.

"6. An elegant valedictory oration was spoken by Mr. Potts, 'On the advantages derived to the study of Physick, from a previous liberal education in the other sciences, particularly Mathematics and Natural Philosophy.'

"7. The Provost then delivered a brief account of the present state of the College, and its quick progress in the various extensive establishments which it hath already made. He pointed out the more general causes of the advancement as well as decline of literature in different nations of the world; and observed to the Graduates, that as they were the first who had received Medical Honours in America, on a regular collegiate plan, it depended greatly on their future conduct and eminence to place such Honours in estimation among their countrymen; concluding with a pathetic exhortation never to neglect the opportunities which their practice would give them (perhaps above all other men), of making serious impressions on their patients, and of exerting the man of Piety and consolation (especially at the awful approach of death), which could not fail to have singular weight from a lay character. He added that what might further concern their practice, he had devolved on a Gentleman of their own profession, from whom it would come with greater propriety and advantage. Upon which Dr. Shippen, Professor of Anatomy and Surgery, gave the remainder of the charge; further animating to graduates to support the dignity of their profession, by a laudable perseverance in their studies, and by a practice becoming the character of gentlemen, adding many useful precepts respecting their conduct towards their patients, charity towards the poor, humanity to all, and the opportunity they might have of gaining the confidence of the sick, and the esteem of every one, who by their vigilance and skill should be restored to health. 'The Vice-Provost concluded with prayer and thanksgiving, and the whole was honoured with the presence of a polite and learned assembly, who by their kind approbation, testified the satisfaction which the inhabitants of this place have in the improvement of useful knowledge in their native country.'"

The Chair of Chemistry was added in 1769, and filled by the selection of Dr. Benjamin Rush. By this appointment, the different professorships originally intended in the medical school were filled up.

At the commencement, held on the 28th of June, 1771, the

The Early History of Medicine in Philadelphia.

first degree of Doctor of Medicine was conferred upon four
of the gentlemen who had graduated as Bachelors in 1768,
viz., Elmer, Potts, Tilton, and Way. On this occasion, Dr.
Morgan delivered the address, in which he entered into a particular account of the branches of study which the Medical
Gentlemen ought to prosecute with unremitted diligence, if
they wished to become eminent in their profession, laying down
some useful rules for an honourable practice in the discharge
of it. He observed that, "The oath which was prescribed by
Hippocrates to his disciples has been generally adopted in
Universities and Schools of Physic on the like occasions, but
that laying aside the forms of oaths, this College, which is of
a free spirit, wished only to bind its sons and graduates by the
ties of honour and gratitude." He therefore begged leave to
impress it upon those who had received the distinguished
degree of Doctor "that as they were among the foremost
sons of the Institution, and the birthday of Medical honours
in this new world, had rose upon them with auspicious lustre,
they would in all their practice, consult the safety of their
patients, the good of their community, and the dignity of
their profession, so as that the seminary from which they
derived their titles in Physic, might never have cause to be
ashamed of them."

(176)

In November, 1779, the College, which had been originally
founded and endowed by private liberality, was by an uncon-
stitutional Act of our Legislature deprived of its charter and
property, which were transferred to a new institution, called
the University of Pennsylvania. This institution was further
enriched by bestowing upon it other property confiscated
during the Revolutionary War. By this Act, the medical
professors, as well as the other officers of the College, lost
their places, and it became necessary to reorganize the medical faculty. In accomplishing this object there was much
difficulty, and a succession of appointments and resignations
took place until the year 1783, when the former professors
were reinstated in the chairs which they had previously held.

In 1789, the Legislature of Pennsylvania, conceiving the
Act of 1779 to be "repugnant to justice, a violation of the
constitution of this commonwealth, and dangerous in its pre-
cedent to all incorporated bodies, and to the rights and fran-
chises thereof," passed a law repealing that part of it which
affected the Academy and College, and thus re-possessed them
of all their former estates and priviledges, but still leaving
the University in existence with its endowments obtained
from the confiscated estates. Thus, two institutions were

established under the titles of College of Philadelphia and the University of Pennsylvania, each having the priviledge of connecting with it a Medical school, and two faculties were soon formed. The College possessed Shippen, Rush, Griffitts, Wistar, and Barton, while the University was sustained by only three professors, Shippen, Kuhn, and Hutchinson. Dr. Morgan, the Founder and first Professor in the school, died before the two faculties were appointed, and to Dr. Shippen the chair of Anatomy and Surgery was given in both institutions.

The unwise policy of attempting to support two schools was so strikingly apparent, that the idea of uniting them was soon thought of, conferences took place, and with the approbation of both parties, application was made to the Legislature to consolidate them into one, to be known as "The University of Pennsylvania," and a law to this effect was passed in September, 1791. The new faculty was fully organized on the 23d of January, 1792, as follows :—

1. Anatomy, Surgery, and Midwifery . Dr. WILLIAM SHIPPEN, JR., and
                                              Dr. CASPAR WISTAR, Adjunct.
2. Practice of Physic . . . . Dr. ADAM KUHN.
3. Institutes and Clinical Medicine . Dr. BENJAMIN RUSH.

4. Chemistry . . . . . . Dr. JAMES HUTCHINSON.

5. Materia Medica . . . . . Dr. SAMUEL POWELL GRIFFITTS.

6. Botany and Natural History . . Dr. BENJAMIN S. BARTON.

Attendance on the lectures of the last-mentioned professors was not essential to graduation.

In the year 1793, the chair of Chemistry was made vacant by the death of Dr. Hutchinson, and Dr. James Woodhouse succeeded to it.

In 1796, Dr. Griffitts resigned the chair of Materia Medica, and Dr. Barton was elected to it.

In 1797, Dr. Kuhn relinquished his Professorship, after having been connected with the school for nearly thirty years, and Dr. Rush succeeded to his duties, though he was not regularly appointed to the place till the year 1805, when the two Professorships of the Institutes and Clinical Medicine, and of Practice were consolidated into one.

In the same year an important change was made in the school by the separation of Surgery from Anatomy and Midwifery, and raising it to a distinct Professorship. Dr. Philip Syng Physick was placed in this chair, and soon afterwards, at his request, in consequence of feeble health, Dr. Dorsey was appointed his adjunct.

The Early History of Medicine in Philadelphia.

In 1810, another great improvement was made by separating Midwifery from Anatomy, and giving to that branch a distinct Professorship. This change was brought about principally by Dr. Wistar, who, in the month of January, 1809, soon after he succeeded to the chair of Dr. Shippen, urged (in a written communication) the Trustees of the University to have Obstetrics separately taught in the school. Early in the following year a resolution to that effect passed the Board, and in the month of June, Dr. Thomas C. James was elected to the new professorship, the first in the country. It was not, however, until the year 1813 that Midwifery was placed on a footing with the other branches taught in the schools and attendance on its lectures rendered essential for the obtaining of a degree. Since that period, other important changes have taken place in the organization of the school—the separation of the Institutes from the chair of Practice in 1835, the addition of a Professorship of Clinical Surgery in 1847, and the prolongation of the lecture term.

May the course pursued by this time-honoured institution still be, as it has been from her foundation, onwards for improvement, and may our science in the Institutions which the prophetic eye of Morgan saw would be everywhere springing up on this Continent, ever be taught with a zeal, fidelity, and dignity, similar to that displayed by her distinguished founder and early professors!

Scanty as is the information which I have been enabled to collect in regard to the early Medical History of our city, it is yet sufficient to show the standing, abilities, and learning of those who practised medicine here from the time of its first settlement down to the period of the Revolution. Themselves scholars, they were the friends of literature and science, distinguished for their liberality and public spirit, and were foremost among those who contributed to the foundation of our institutions of learning, benevolence, and science. Almost forgotten as their labours now are, it has been a pleasing task to recall their virtues and activity. Of such men our profession must feel justly proud. When we consider that it was but little more than forty years before the institution of our medical school that the celebrated one of Edinburg was first formed, and that at the period of the foundation of our Hospital, but five similar institutions existed in the metropolis of Great Britain, in one of which the instruction of pupils was not permitted by the Governors, while in another of them but nine at a time were admitted, and that Clinical Medicine was not cultivated in France, until a short period prior to their Revolution; it will be evident that the teachers of our school and hospital, and the practitioners who encouraged and supported them, were not only awake to the advance and improvements in the healing art, but also quick to adopt the good practices of Europe.*

* In 1725 there were but three hospitals in the city of London for the sick and lame, St. Bartholomew's, St. Thomas's, and Guy's. The latter was founded only in 1722. St. George's was founded in 1733. The London Hospital in 1740; and the Middlesex in 1750. Until 1729, when the Surgeons and Physicians opened a house for the reception of poor patients, no hospital existed in Edinburg, and the Royal Infirmary there was not founded until the year 1736.*

The first Clinical Schools were established in Italy about the middle of the sixteenth century; but there appears to have been no clinical teaching of much note till the time of Boerhaave, who acquired great renown by it. In 1753, Van Swieten opened a clinical hospital at Vienna, and was followed by De Haen, Stoll, and Hildenbrand. Dr. Rutherford introduced the system to Edinburg, and is stated to have been the first who gave clinical instruction in Great Britain. He was succeeded by Cullen and others. In France, it was not till the year 1794 that there was any clinical organization, when Desault and Corvisart were made the first professors. Dr. Taylor† states that clinical lectures were not delivered in London until the commencement of the present century. After their introduction

---

* Spence, Edin. Med. Journ., vol. 10, page 489.

† Introductory Lecture, Lancet, 1841-2, page 47.

May their names ever be fondly cherished by their successors, and their many noble qualities and respect for learning be perpetuated among us !

---

they appear to have been discontinued, for Dr. Billing informs us "that at the time he adopted the practice in 1822, there were none given in London."

Otley, in his life of Hunter,[*] mentions that up to the time of Dr. William Hunter (1745), lecturers on Anatomy in Great Britain "had been accustomed to employ but one subject for demonstrating all parts of the body, excepting the bones and arteries, which were described on preparations; and the nerves, for exhibiting which, a fœtus was usually employed. Practical dissection was unknown to the great bulk of the profession. The lecturers of that time," says he, "treated in one course on a number of subjects sufficient to furnish matter for three or four distinct courses, according to our present system." Mr. Bromfield, a teacher of considerable note, comprised Anatomy and Surgery in a course of thirty-six lectures; and Dr. Nicholls, in whose school Dr. William Hunter studied, "taught Anatomy, Physiology, and general principles of Pathology and Midwifery in thirty-nine."

---

[*] The Works of John Hunter, by James F. Palmer, London, 1835, vol. 1, pp. 5 and 6.

The End.

Appendix No. 1.

ORGANIZATION OF THE MEDICAL DEPARTMENT AND OF GENERAL HOSPITALS.

THE following particulars relating to the organization of our Military Hospital Department, and the early action of Congress in regard to it may prove interesting:—

July 17, 1775. A committee of three was appointed by Congress to report a method of establishing a General Hospital Department. The committee consisted of Messrs. Paine, of Massachusetts, Lewis, of New York, and Middleton, of South Carolina.

July 27. Congress "Resolved, That for the establishment of an Hospital for an army, consisting of 20,000 men, the following officers and other attendants be appointed, with the following allowance and pay.

"A Director-General and Chief Physician, his pay four dollars per day.

24 (185)

Four Surgeons per day, each one and a third dollars.

One Apothecary, one and a third dollars.

Twenty mates, each per day, two-thirds dollar.

One clerk, two-thirds dollar.

Two storekeepers, each four dollars per month.

One nurse to every ten sick, one-fifteenth of a dollar per day, or two dollars per month.

Labourers occasionally."

Resolved, "That the appointment of the four Surgeons and the Apothecary be left to the Director-General and Chief Physician. That the Mates be appointed by the Surgeons, and that the number do not exceed twenty; and that the number be not kept on constant pay, unless the sick and wounded should be so numerous as to require the attendance of twenty, and to be dismissed as circumstances will admit; for which purpose the pay is fixed by the day, that they may only receive pay for actual service. That the Clerk, Storekeepers, and Nurses be appointed by the Director." Dr. Church was unanimously appointed Director-General at the same time. From the estimates given into Congress, it was computed that £10,000 sterling per annum was sufficient to answer all demands of the General Hospital for these 20,000 men, and this when at least one thousand men, in every five, were sometimes considered unfit for duty.

September 14, 1775. Congress "Resolved, That Samuel Stringer, Esq., be appointed Director of the Hospital, and Chief Physician and Surgeon for the army in the Northern Department," pay four dollars per day. "That he be authorized to appoint four Surgeon's mates, who are to receive pay, and be engaged on same terms as ordered by the Resolves of July 27th." "That the Deputy Commissary-General be directed to pay Dr. Stringer for the medicines he has purchased for the use of the army, and that he purchase and forward such other medicines as General Schuyler shall by his warrant direct for the use of the said army."

October 14, 1775. Dr. Church, Surgeon-General of the Army, "having been found guilty of traitorous practices, in corresponding with the enemy," was removed.

October 17, 1775. Dr. John Morgan was elected.

December 8, 1775. One Surgeon allowed to each battalion. Pay $25 per month, and W. Barnet, Jr., was elected Surgeon to Lord Sterling's, or first battalion raised in New Jersey.

June 6, 1776. Resolved, "That Dr. Jonathan Potts be employed as a Physician and Surgeon in the Canada Department, or at Lake George, as the General shall direct, but that this appointment shall not supersede Dr. Stringer."

July 12, 1776. A resolve of Congress forbidding the issuing of Hospital stores to Regimental Surgeons, etc.

July 15, 1776. A Chief Physician appointed for the Flying Camps, and Wm. Shippen, Jr., elected. Pay $4 per day.

July 17, 1776. Congress "Resolved, That the number of Hospital Surgeons and mates be increased in proportion to the augmentation of the army, not exceeding the surgeon and five mates, to every five thousand men, to be reduced when the army is reduced, or when there is no farther occasion for so great a number." Resolved, "That the pay of the Hospital Surgeons be increased one dollar and two-thirds of a dollar per day; the pay of the mates be increased to one dollar per day, and the pay of the apothecary be increased to one dollar and two-thirds of a dollar per day; and that the Hospital Surgeons and mates take rank of Regimental Surgeons and mates."

No Regimental Surgeon to draw on the Hospital for any stores, except medicines, instruments, etc.

July 20, 1776. Congress resolved, "That Dr. Scuters be recommended to Dr. Morgan, who is desired to examine him, and if he find him qualified, to employ him in the Hospital as Surgeon."

August 20, 1776. Dr. Stringer presented a petition to Congress "to request an explanation of the resolves of Congress, respecting the nature and extent of his own as well as Dr. Morgan's appointment."

The committee appointed to consider it, report:—

"That Dr. Morgan was appointed Director-General and Physician-in-Chief of the American Hospital.

"That Dr. Stringer was appointed Director and Physician of the Hospital of the Northern Department only."

They go on to say that "every Director of a Hospital possesses the exclusive right of appointing Surgeons and Hospital officers agreeably to the resolves in his own department." They also recommended the appointment of a general druggist, and Dr. William Smith was appointed.

September 19, 1776. An assistant physician to Dr. Shippen was allowed. Dr. Wm. Brown was elected by Congress.

About this time a resolution was passed by Congress (after Sept. 24, 1776), requesting the several States to appoint skilful Surgeons and Physicians to examine the Surgeons and Surgeon's mates, who offered themselves to serve in the army or navy, and declaring that no commission should be issued to any who should not produce a certificate from such examiners, that they were qualified for the duties of their office.

October 9, 1776. It was resolved "That for the future no Regimental Hospital be allowed in the neighbourhood of the general hospital," and that "John Morgan, Esq., provide and superintend a hospital at a proper distance from the camp, for the army posted on the East Side of Hudson's River, and that William Shippen, Esq., provide and superintend a hospital for the army in the State of New Jersey," and various directions are given respecting their duties, that they make weekly returns to Congress and the Commander-in-Chief, that each of the Hospitals be supplied "by the respective directors, etc."

November 29, 1776. It was resolved, among other things, "That the General in each army cause strict inquiry into the conduct of the directors of the Hospitals and their surgeons, as also of the Regimental Surgeons."

A letter was received by Congress from Dr. Shippen, and they resolve "That Dr. Morgan take care of the sick on the East Side of the Hudson River, and Dr. Shippen those on the West Side."

In November, 1776, a Congressional Committee, who had
been investigating the affairs of the army, and had visited
the General Hospital of the Northern Department at Fort
George, made a damaging report; in consequence of which,
on the 9th of January, 1777, Congress resolved "That Dr.
Morgan, Director-General, and Dr. Stringer, Director of the
Northern Department, be dismissed." To this report, and the
dismissal of Dr. Stringer, General Schuyler, commanding the
Northern Department, took offence, and expressed himself
unguardedly in relation to their action in some of his official
letters. This gave rise to a curious correspondence and alter-
cation between Congress and himself, and on the 15th of
March, the former resolved, "That as Congress proceeded to
the dismission of Dr. Stringer, upon reasons satisfactory to
themselves, General Schuyler ought to have known it to be
his duty to have acquiesced therein. That the suggestion in
General Schuyler's letter to Congress that it was a compliment due to him to have been advised of the reasons of Dr.
Stringer's dismission, is highly derogatory to Congress, and
that the President be desired to acquaint General Schuyler
that it is expected his letters for the future be written in style
more suitable to the dignity of the representative body of

these free and independent States, and to his own character as their officer :"—

Resolved "That it is altogether improper and inconsistent with the dignity of this Congress to interfere in disputes subsisting among the officers of the army, which ought to be settled, unless they can otherwise be accommodated in a court-martial, agreeably to the rules of the army, and that the expression in General Schuyler's letter of the 4th of February, 'That he confidently expected Congress would have done him that justice, which it was in their power to give, and which he humbly conceives they ought to have done,' was to say the least ill-advised and highly indecent."

The General afterwards presented a Memorial to Congress, explaining the expressions in his letter which had given them offence. "In this," he says, "the power of Congress to dismiss their servants without a formal inquiry, your memorialist, for his own part, never questioned; but its policy, as a general rule, he humbly begs leave to observe, may be subject at least to one strong objection: it may tend to prevent men of worth and abilities from affording to the public that assistance which they are capable of giving, from the apprehension that the suggestion of clamours, too often arising from a jealousy of office, might expose them to disgrace and injury of a dismission, without being heard in their own defence;" and further on he says, "that he took it for granted that Congress was acquainted, that he had in a manner forced Dr. Stringer in the service; that in August, 1775, when sickness was spreading through the army under his command with great rapidity, and they were not only destitute of competent medical assistance, but even of medicines, his repeated solicitations, supported by the promises of a Member of Congress (the late Mr. Lynch), prevailed on Dr. Stringer to exchange an extensive and well-established practice for your service, and to appropriate a large stock of his own medicines to the public use." He further begged "leave to observe that Dr. Stringer, since his dismission, without any inquiry into his conduct, imputes the loss of a profitable business, as well as that of his medicines, which cannot now be replaced, to your memorialist, who, for that reason, could not be anxious to have it in his power to assign the motives to Congress for taking the measure," and that he "had expressed his wish of being informed of the reasons for dismissing Dr. Stringer, not as a right, but merely as a matter of compliment, and not from impatience and curiosity, but with a view to obviate that gentleman's complaints," and that "he did not mean to wound their dignity, or dispute their authority." This explanation of the expression in his letter which had given them offence seems to have satisfied Congress, and on the 8th of May they resolved "that the explanation was satisfactory, and that now they entertained the same favourable sentiments concerning him, which they entertained before that letter was received." A fortnight after this resolution, General Schuyler was directed again to take command of the Northern Department of the army.

Appendix No. 2.

SCANT SUPPLIES AND SUFFERINGS OF THE
REVOLUTIONARY ARMY.

THE scant way in which the army was provided in the early years of the war, and their urgent wants during the whole course of it, may be judged of from the following.

In 1776 a general order was issued from headquarters to the Regimental Surgeons, to report to the Director-General the instruments, bandages, etc., they had on hand in order to lay them before the Medical Committee of Congress. Reports were received from fifteen regiments, by which it appeared that for fifteen Regimental Surgeons and as many mates, all the instruments, and they were private property, were six sets for amputations, two for trepanning, fifteen cases for the pocket, seventy-five crooked, and six straight needles. Among the whole fifteen surgeons, there were only four scalpels, three pair of bullet forceps, half a paper and seventy pins, but few bandages, ligatures, or tourniquets, and only two ounces of sponge. Dr. Morgan, in his report to the Committee of Congress, observes that he "supposes the reporting regiments to be at least as well provided as any others that have neglected to pay due attention to the order."

On one occasion, a gentleman belonging to the General Hospital, Dr. Binney, was sent from the army in New York to Philadelphia to purchase instruments, if to be had ready made, or to employ workmen to make them, but it was found "that there were no instruments to be purchased at any rate, and that the only workman in the city that could make surgeon's instruments was engaged by Congress upon arms, and could not undertake any work for a long time to come."

In June, 1777, a medical officer at Trenton wrote to Dr. Potts, at that time Director of the Military Hospital at Philadelphia, as follows: "The amputating instruments which you sent for the use of the Hessian surgeon were taken away yesterday by Mr. Wood, and I am told by the former that they were so bad that he could not make use of them, had they been left. There are four or five men who must submit to the operation soon or lose their lives, I therefore beg you would, with the medicines, also send up a compleat set."

Dr. Hutchinson, in one of his letters, asserts "that during the winter of 1778 there were such a want of lancets, that numbers of the Regimental Surgeons, and some of those of the Flying Hospital were without one."

The following extracts from letters will also strikingly make known the great wants and necessities of the army in all things.

DR. POTTS TO DR. MORGAN.

FORT GEORGE, August 10th, 1776.

The distressed situation of the sick here is not to be described, without clothing, without bedding, or a shelter sufficient to screen them from the weather. I am sure your known humanity will be affected, when I tell you we have at present upwards of 1000 sick, and crowded into sheds, and labouring under the various and cruel disorders of Dysenteries, Bilious Putrid Fevers, and the effects of a confluent smallpox; to attend this large number, we have four seniors and four mates, exclusive of myself, and our little shop doth not afford a grain of Jalap, Ipecac., Bark, Salts, Opium, and sundry other capital articles, and nothing of the kind is to be had in this quarter; in this dilemma, our inventions are exhausted for succedaneums, but we shall go on doing the best we can, in hopes of a speedy supply. Dr. Stringer left us some days since in order to lay the situation of the Hospital before his Excellency General Washington, and endeavour to procure redress. . . . . Dr. Stringer and myself have had some conversation respecting the expediency of acting under a Director-General of the whole Continent; this the Dr. was averse to, and mentioned some reasons which had weight with me; as you will see the Dr., I need not take up your time in mentioning them. For my own part, I am resolved to be governed by such regulations as our wise Congress shall think proper, wishing nothing more than to contribute my mite towards the relief of our once distressed country, but now the glorious independent States of America. Pray present my most respectful compliments to his Excellency Gen. Washington, and to Gen. Mifflin, and believe me to be, dear sir,

Your affectionate and most humble servant,

JON'N POTTS.

DR. STRINGER TO DR. POTTS.

FORT GEORGE, Oct. 29th, 1776.

DEAR POTTY :—

Mr. Stockton, one of the members of Congress, has been here, and inquiring for a bear-skin learned that you were master of one, which was then said to be on this ground. I promised to endeavour to procure it, or another for him, as he was very anxious to get one, and yesterday got a letter from him about it. If you can spare it, you will much oblige him; he mentions a desire to know the price of it. I shall be glad if you would let him have it, and give me word as soon as possible. Don't grumble about it, Potty, like the animal that first wore it. The nurses are wanting their wages. My dear friend, I wish it was in my power to send you the non-naturals you request. The shrub is entirely out, but little spirits, no vegetables yet to be had. Sugar almost expended, and a great concourse of friends to entertain daily (Militia) will throw us soon to want. Brown mends but slowly. My compliments to ———, and I am,

Yours, sincerely,

SAMUEL STRINGER.

(199)

HEADQUARTERS, July 26th, 1777.

DEAR SIR:—

Your favour of the 25th inst. I have this moment received. Shall comply with the requisition contained, though I shall be left with but two Regimental Surgeons in the whole army. I am this moment returned from Fort Edward, where a party of Hell Hounds, in conjunction with their brethren, the British troops, fell on our advance guard, and inhumanly butcherd, scalped, and stripd four of them, and wounded two more, each in the thigh; four more are missing. Poor Miss Jenny McCrea, and the woman with whom she lived, were taken by the savages, led up the hill to where there was a body of British troops; there the poor girl was shot to death in cold blood, scalped, and left on the ground; the other woman not yet found. The alarm came to camp at two P. M.; I was at dinner. I immediately sent off to collect all the Regimental Surgeons, in order to take some one or two of them along with me to assist, but the Divil a bit of one was there to be found, except three mates, one of whom had the squirts; the other two I took with me. There is neither amputating instruments, crooked needle, or tourniquet in all the camp. I have a handful of lint, and two or three bandages, and that is all. What in the name of wonder I am to do in case of an attack, God only knows; without assistance, without instruments, without everything. What can become of Stewart with the stores, medicine chest, my baggage, etc. If it is consistent with the public good, and agreeable to your opinion, pray assist me with one or two of your surgeons. My respectful compliments to yourself and all the Fraternity.   I am, Sir,

Your very humble servant,

DR. POTTS, Dep. Director-Genl. N. D.   JNO. BARTLETT.

DR. JOHN COCHRAN TO DR. POTTS, MORRISTOWN.

DEAR SIR:—

I received your favour by Dr. Bond, and am extremely sorry for the present situation of the Hospital finances; the stores have been all expended for two weeks past, and not less than 600 Regimental sick and lame, most of whom require some assistance, which being withheld, are languishing, and must suffer. I flatter myself you have no blame in this matter, but curse on him or them by whom this evil is produced. The vengeance of an offended Deity must overtake the miscreants sooner or later. It grieves my soul to see the poor, worthy, brave fellows pine away for want of a few comforts which they have dearly earned. I shall wait on his Excellency, the Commander-in-Chief, and represent our situation, but I am persuaded it can have little effect, for what can he do? He may refer the matter to Congress, they to the Medical Committee, who would probably pow-wow over it for a while, and no more be heard of it. Thus we go before the wind. Compliments to all friends, and believe me, Dear Sir,

Yours, very sincerely,

JOHN COCHRAN.

THOS. BOND, Jr., TO DR. POTTS, MORRISTOWN.

DEAR POTTS:—

The Hospitals are all suffering for want of stores, particularly wine, spirits, tea, coffee, rice, and molasses. Very grievous complaints were made me yesterday by Dr. Moses Scott and Tilton, at Baskenridge; they suffer very much by the commissioners not being furnished with cash by which they can procure milk and vegetables, matters so necessary to a sick person. God! 'twould make you feel and rouse every pulse within you to see a fine brave fellow who has nobly fought in most of our battles, perhaps been dangerously wounded in one or more, and by the application of some prudent and generous remedies which were in our power then to furnish him with, soon recovered; I say would rouse every feeling now to see this brave man languishing on a sick bed, with his physician holding his wrist, and promising to send him some more *Physic*, when perhaps a glass of generous wine, or some comfortable hospital store would rouse his drooping spirits, and prolong that life which has, and is from principle devoted to the service of his country. I shall talk the matter over very freely with the General the first opportunity, and let him know our situation, by which the blame must be taken off your shoulders. I have wrote a small tale of the pathetic for our brave soldiers, but I have another grievous one to relate for ourselves. Joe and myself have spent all our money, and fear, unless we can borrow, we shall starve; do pray prevent it by sending us cash. You may depend upon it, no Surgeons of the army can lend us a shilling. Yours,

THOS. BOND, Jr.

Dr. Jonathan Potts, Dep. Director-Genl., Philadelphia.

The Early History of Medicine in Philadelphia.

CAMP SINIAH, September 1st, 1780.

SIR :—

Mrs. Schuyler has informed me that she lent you a hair mattress, covered with yellow-striped furniture check, when you were in Albany, in 1777, which has not been returned to her. You will please to send it, or one like it, to Col. Hamilton, Aid-de-Camp to his Excellency, the Commander-in-Chief.

I am, Sir, your most obedient,

Humble servant,

TO JONATHAN POTTS, ESQ.

PH. SCHUYLER.

Appendix No. 3.

DR. MORGAN'S CARD INVITING PUBLIC INVESTIGATION OF HIS CONDUCT.

Boston, April 10th, 1777.

"On Monday arrived Dr. John Morgan, late Director-General of the Hospitals, and Physician-in-Chief to the Continental Army. He has been informed that several evil-minded persons have taken upon them to circulate a number of false and groundless reports in this place, with a view to injure his reputation in that public station, and amongst other particulars, has learned that loud clamours have been industriously raised against his having taken possession of the medicines, shop furniture, and laboratory of Dr. Sylvester Gardiner* and

---

* Dr. Gardiner was the most noted and extensive druggist in New England, and took sides with the mother country. He escaped to England, and the Legislature of Massachusetts having enacted that all property belonging to Tory refugees should be confiscated for public use, his estate was sold at public auction. His stock of drugs was said to have filled over 20 wagons.—THACHER.

Dr. William Perkins of this town for the use of the army; and likewise, that many of the sufferings of the sick in the last campaign arose from his having unjustly withheld from them (or from their Regimental Surgeons) those stores which they were entitled to draw from the General Hospital, suggesting, in the first instance, that his motives for so doing was his private emolument in the latter, to oblige the sick to be sent to the General Hospital, that he might stop their rations whilst there, and put the money in his pocket. As these reports do not merely affect private character, but concerns the public to know the truth thereof, he hopes for their indulgence in making it known in a newspaper, that he has called on the persons pointed out to him for more particular information of the grounds of these reports, and means to lay the matter of his inquiry before the public.

"That the necessary information to clear up this matter may not be kept back, he does hereby publickly advertise, that if there are any persons who can pretend to a knowledge of these facts, who will take upon them to answer to the charge, he now invites them to step forth and state their accusation against him in either of these particulars, or any other that regards the faithful discharge of his public trust, so that the said facts may be properly ascertained and examined. In return, he undertakes to lay before the public a faithful and exact account of every proceeding they require relative to the discharge of his duty in the above station, by which the world will be enabled to judge whether any of the charges are well founded, or only proceed from a spirit of malignity and detraction."—*Towne's Penn. Even'g Post, No. 342, for April 22d, 1777.*

Appendix No. 4.

MEDICINE IN THE SISTER COLONIES.

DURING the period of which we have been speaking medicine was cultivated with more success in New York and some of the Southern States than in New England. In the South the study of Natural History was already pursued with zeal, and the close affinity of many of the branches of physical science to medicine, with its tendency to attract the mind to observations in anatomy and to the physiological and pathological state of the system, are all well known. In these States, too, not a few of the physicians were of European birth, and were possessed of easy fortunes, and well-selected libraries, with eager and cultivated minds, who naturally infused their own enthusiasm into those around them, and created a standard of excellence superior to what existed elsewhere. The greater wealth of these States also allowed their citizens to confer on their sons the benefits of an European education.

Of elaborate medical writing, which even yet may be said to be in its infancy with us, we possessed little or none in the last century, and, until within a comparatively short time, our medical literature consisted principally in brief memoirs on particular topics, practical details, or uncommon cases. This may be readily accounted for. The most eminent of the physicians of those days, toilsomely engaged in the practice of their profession, were prevented both by want of time and of facilities from wooing literature, while the greater portion of them, from want of erudition, classical attainments, and that necessary discipline of mind which the study of these produces, were perhaps incompetent of any literary efforts.

The first medical publication in the country of which we have any account was a paper in 1677 by Thomas Thatcher, a clergyman and physician, entitled "A Brief Guide to the Smallpox and Measles." In 1721, Zabdiel Boylston, celebrated as the introducer of inoculation among us, published in the Philosophical Transactions an account of inoculation by Timonius, of Constantinople, and Pylarinus, a Venetian Consul in Smyrna, and in 1726 he furnished an historical account of inoculated smallpox in New England.

A Scotch physician, named Douglass, who emigrated to

27

Boston, in 1720, is the next author of whom any account has reached us. He was a violent opposer of the practice of inoculation, and wrote essays concerning the smallpox in 1722 and 1730. In 1736 he published "The Practical History of a New Epidemical, Eruptive Miliary Fever, with Angina Ulcusculosa, which prevailed in New England in 1735, and '36," which has been reprinted in the fourteenth volume of the New England Medical and Surgical Journal, and is a valuable literary production. To Dr. Douglass is due the high honour of introducing the use of mercury as a remedy in the treatment of acute inflammatory diseases, the first suggestion of which will be found in the essay just named. Another work of this author was "A Summary, Historical and Political, of the First Settlement, Progressive Improvements and Present State of the British Settlements in North America," in two volumes, which attracted much attention, and reached a second edition; the first was published in Boston, 1749, and the second in London, 1760. In it will be found frequent remarks concerning the state of medicine in the colonies. In 1732 a tract on Pharmacy was written by Thomas Harward, a clergyman; and in 1742 a pamphlet on the method of practice in the smallpox was published by Nathaniel Williams, who is represented as a learned physician, who enjoyed an extensive practice for thirty-seven years, and, in addition, was an instructor of youth, a preacher, and most valuable citizen.* With the exception of the works of Boylston, and of Douglass, these publications were of little account, and their existence at this time is almost forgotten.

In addition to these, the names of Benjamin Gale and William Hunter deserve mention. The first was a distinguished practitioner of medicine in Connecticut, and published in 1740 a "Dissertation on the Inoculation of the Smallpox in America." This was intended as a prize dissertation for the Academy of Bordeaux, and in it he advocated the utility of a course of mercury as a preparation to that disease, as recommended by Dr. Thompson. It was published in the Philosophical Transactions (q. vid. vol. 12, abridged, p. 229), and did him great credit both in this country and in Europe.

Hunter was a native of Scotland, and settled in Rhode Island about the year 1752. He was a pupil of Monro the elder, and gave lectures at Newport on the History of Anatomy and Comparative Anatomy in 1754, '5 and '6, the

* Bartlett, Mass. Med. Coun., vol. 2, p. 239.

first ever delivered on the science in America. He died in 1777.

The account given of the profession in early times, in New York, by Smith, the historian of that State, is not a favourable one. "Few physicians," says he, "are eminent for their skill. Quacks abound like locusts in Egypt, and too many have recommended themselves to a full practice and profitable subsistence." (History of New York from its First Discovery to the Year 1732, Lond. 1757.) From the statements of Dr. Douglass too (q. vid.) it would appear that the profession there was less elevated in its character than in the other principal towns of our continent. This applies, however, only to the first part of the last century, as towards the middle of it we find several individuals worthy of a distinguished place in the Annals of American Medicine as having eminently contributed by their writings to its advancement.

Among the most celebrated of these was Cadwallader Colden. This gentleman emigrated to that province from Scotland in 1718, and though he soon relinquished the practice of his profession, yet he ever continued to feel a lively interest in medical pursuits. To the study of botany he was ardently devoted, and described between three and four hundred American plants, which were afterwards printed in the Acta Upsaliensa (Beck's Hist. Sketch). His medical writings consist in "An Account of the Diseases and Climate of New York," published about 1720 (Med. and Philosoph. Reg., vol. I.), "Observations on Fever which prevailed in the City of New York in 1741, '42" (Med. and Philosoph. Reg., vol. I.), and a letter to be found in the London Med. Observs. and Inquiries, upon the "Sore Throat Distemper" which prevailed epidemically in 1735. He entered into public life, and held in succession the offices of Surveyor-General of the Province, Master in Chancery, member of the Council, and Lieutenant-Governor.

Another prominent man of this period was Dr. John Bard. He prosecuted his medical studies under Dr. Kearsley, of Philadelphia; and after practising some years in that city he removed to New York. He was the author of an "Essay on the Nature and Cause of Malignant Pleurisy," which had prevailed extensively on Long Island in 1749 (Med. and Phil. Regist., vol. I.), and furnished for the London Med. Obs. and Inq. for 1760 "A Case of Extra-Uterine Fœtus;" besides several papers on the Yellow Fever. In connection with Dr. Middleton he dissected in 1750 the body of an executed murderer and injected the vessels, for the instruction of their students. This is generally stated to have been the first dissection made in America, but we have, in a former part of this paper, shown that so early as 1742 Dr. Cadwalader made dissections in Philadelphia, for the benefit of his brother practitioners. Dr. Bard died in 1799, aged 83. Dr. Jacob Ogden, a native of New Jersey, but settled on Long Island, published in 1769 and 1774 letters on the "Malignant Sore Throat Distemper." The credit of being the first to use mercury in the treatment of fevers and acute inflammatory affections in the United States has by many been given to Dr. Ogden. This, however, is inaccurate. Dr. Douglass, of Boston, had used it extensively in the Angina Maligna as far back as the year 1736 (his papers in N. E. Journal, vol. 14, p. 4), and Dr. Holyoke, of Salem, as early as 1751, pursued the same practice (Thacher).*

Samuel Bard was born in Philadelphia in 1742. He received

---

* Dr. Warren, in his Treatise on the Malignant Fever of Barbadoes, published in 1740, animadverts on the practice of giving calomel in inflammatory fevers, thus showing that, previous to this time, the remedy had been in use. See Francis' account. See also Hamilton in Duncan's Commentaries for 1785, and the *Résumé* in Johnson's Review, May, 1828, vol. 9, page 238.

his medical education, and graduated at Edinburg in 1765. Within a year after his return to New York a medical school was organized and connected with King's College in that city, in which he was appointed Professor of the Theory and Practice of Medicine. In his address at the commencement in 1769 he enforced the necessity of a public hospital, and to his benevolent exertions, in great measure, is New York indebted for the foundation of her City Hospital. He was highly eminent as a practitioner, and for very many years stood among the first in his profession. In 1771 he published an essay on "Angina Suffocation," and this, together with a paper upon the use of Cold in Hæmorrhage, and a treatise upon Obstetrics, published at a much later period, are his principal publications. (McVickar's Eulogium, p. 187.)

Of the medical men of our sister city, however, none are more worthy of distinction than Peter Middleton, and Richard Bayley. Dr. Middleton was a native of Scotland, and was distinguished for his learning and great professional talents; he was among the most active in the foundation of the New York Medical School, in which institution he ably filled the professorship of Physiology and Pathology, and also contributed to the establishment of the Hospital. He was the author of an "Historical Inquiry into the Ancient and Present State of Medicine," the substance of which was delivered at the opening of the Medical School in 1769. A copy of this is in possession of Dr. Beck, who mentions it as exhibiting an extensive survey of the state of medicine among the different nations of the globe, and as affording ample proof of the learning and ability of its author. In 1780 he published a letter on Croup, addressed to Dr. Bayley, in which he sanctions and confirms the propriety of the practice which had been recommended by that writer. He died in 1781.

Richard Bayley was born in 1745, and completed his medical education in London with the Hunters. He was distinguished as a surgeon; was an experienced and successful lithotomist, and well skilled in ophthalmic surgery. As early as 1782 he amputated at the shoulder-joint successfully, which is believed to have been the first instance of its having been done in the United States. In 1781 he published "Cases of Angina Trachealis," with the mode of cure in a letter to William Hunter, M.D., which is worthy of particular attention. In this tract he clearly established the fact of the disease being inflammatory in its nature, and of the false membrane and effusion being the mere consequences of it. By a comparison of its pathological appearances he also distinguished the disease from Angina Maligna with which up to this time it was generally confounded, and in its treatment he urges that decided and energetic course which has been confirmed by all subsequent experience, viz., repeated bleedings, the free use of tartar emetic, and other evacuants, with blisters over the larynx. Though not published until 1781, yet the views of the disease and its treatment held by the author had been made known to many, and had been published by his friend, the celebrated Michaelis, in Richter's Surgical Repository (Watson, p. 8). In 1787 he delivered a course of lectures on Surgery, and in 1792 was elected to the Professorship of Anatomy in Columbia College. In 1796 he published an account of the epidemic fever which prevailed in New York during the previous year. He died in 1801.

John Jones was a native of Long Island, and, after completing his medical studies in Philadelphia, improved himself by a visit to Europe, and, upon his return, devoted himself particularly to the practice of surgery. He was the first to perform the operation of lithotomy (about 1753) in New York City, and his success was such in several cases, which soon presented themselves to him, that he became well known as an operator throughout the Middle and Eastern States. (Mease, p. 10.) Upon the foundation of the New York Medical School, he was appointed to the professorship of Surgery. At the commencement of our Revolutionary war he published his "Plain Remarks upon Wounds and Fractures," a work intended principally as a guide for young surgeons of our army and navy in the classes of accidents to which their attention then was continually directed. This book, which embodies the sentiments of the best surgeons of the period, on the subjects treated of, with the results of the author's own observations, contains much valuable matter, and well put together. It passed through three editions; the first was published at New York in 1775, and the two latter at Philadelphia in 1776 and 1795. He settled here in 1780, and was in the following year elected one of the surgeons of our Hospital, and upon the foundation of the College of Physicians, in which he took a prominent part, was made one of its Vice-Presidents. He died in 1791.

Charles McKnight was born in New Jersey in 1750. After completing his collegiate education at Princeton he came to Philadelphia, and commenced the study of medicine under Dr. Shippen. He afterwards entered the American army, and at the termination of the war settled in New York, and delivered lectures on Surgery and Anatomy. As a practitioner of Surgery he was much distinguished; and, excepting Dr. Bayley, was without a rival in that branch in New York. His only published production is a case of "Extra-Uterine Fœtus," in the mems. of the Medical Society of London, vol. 4. He died in 1791, aged 41.

In Maryland, two Scotch physicians, Drs. Hamilton and Adam Thomson, practised many years with distinction. The former was the preceptor of Dr. Bond. The latter, after some years, removed to Philadelphia, where, in 1750, he published "A Discourse on the Preparation of the Body for the Reception of the Smallpox." This production received much praise both at home and abroad, as at that period inoculation was on the decline. The author states that the practice was so unsuccessful at Philadelphia, that many were disposed to abandon it; wherefore, upon the suggestion of the 1392d aphorism of Boerhaave, he was led to prepare his patients by a composition of Antimony and Mercury, which he had constantly employed for twelve years, with uninterrupted success. (Watson's Annals of Philadelphia, vol. 2, page 376.)

At the period of which we are now speaking, Virginia possessed some professional men of merit. In 1736, Dr. John Tennent, a native of that State, communicated to the Edinburg Medical Essays and Observations a paper recommending the Polygala Senega in pleurisy and peripneumonia, which he had used with much success after depletion.

Dr. John Mitchell, who emigrated to Virginia from England, at the beginning of the last century, was also distinguished as a physician, and a man of learning. To him we are indebted for a well described account of the yellow fever, which prevailed in Virginia in 1741 and '42. This fell into the hands of Dr. Colden, of New York, by whom it was sent to Dr. Rush, and has been published in the Medical Museum, vol. 1. In the treatment of this disease he depended principally upon diaphoretics and purgatives, and Dr. Rush has acknowledged that it was from his account that he was led to the free use of cathartics in that disease, and that he had derived from it much else that was highly useful to him, in directing the treatment of the disastrous fever of 1793.

In 1769 Dr. Mitchell published a quarto volume on the general principles of botany, and furnished descriptions of a number of new genera of plants. This, and an elaborate essay "On the Causes of the Different Colours of People in Different Climates," which was published in the Philosophical Transactions (vol. 43, 17), were his principal scientific publications.

Another physician of Virginia, well known for his botanical knowledge, was John Clayton. Like Mitchell, he was an Englishman, and emigrated to our country in 1705 (Thacher, 224). In addition to several papers communicated to the Philosophical Transactions upon medicinal plants, he published in 1743 the "Flora Virginica," a book which was deemed worthy of republication at Leyden, in 1739 (see Pennsylvania Hospital Catalogue).

Previous to the time of Rush, no men on our Continent were more celebrated for their observations and writings than those of Carolina (see Ramsay's History of South Carolina). In 1743 Dr. John Lining contributed to the Transactions of the Royal Society of London (vol. 42, p. 491, Thomson's History of Royal Society, p. 129) a paper containing a series of statical experiments performed upon himself, perhaps the most valuable that have ever been published. Throughout the whole of the year 1740 he carefully continued these experiments, ascertaining his weight each morning and evening, together with the weight of his food, and of his urinary and alvine discharges, and in 1753 he published in the Edinburg Essays and Observations (vol. 2) an accurate account of the "American Yellow Fever," the first that has been given to the public from our Continent, and remarkable for its accuracy and minuteness of detail. Dr. Lionel Chalmers, another physician of Charleston, communicated to the Medical Society of London, in 1754, a good paper on "The Opisthotonos and Tetanus" (q. vid.), which was published in the first volume of their Transactions. In this essay he recommends venesection, the warm bath, and free use of opium and emollient enemata, as the remedies principally to be relied upon. In 1767 he published "An Essay on Fevers," the origin of which he endeavours to show is not to be looked for in the state of all fluids, but in the solids, and holds their immediate cause to be a "spasm of the arteries and other muscular membranes," a doctrine which had before been glanced at by Hoffman, but which it was reserved for Cullen afterwards completely to illustrate, and which served as the foundation of much of his great fame. In addition to these, Dr. Chalmers published in London, in 1776, a valuable work in two volumes upon the diseases and weather of South Carolina. In 1764 Dr. Alexander Garden gave an account of the Spigelia Marilandica and its virtues (Edinburg Essays and Observations, vol. 3), besides communicating to the Royal Society several interesting papers on natural history, to the study of which he was ardently devoted.

The three eminent men we have just mentioned had all emigrated from Scotland, at an early period of the last century, and their observations and writings at this early period, in addition to the learning, great practical skill, and virtues of Bull, the Moultries, and Ramsay, have justly rendered the profession in Charleston distinguished in our medical history.

Appendix No. 5.

SOME PHILADELPHIA PHYSICIANS OF LATER DATE.

Adam Kuhn was born in Germantown, Pa., in 1741, and read physic with his father. He went to Upsal in 1761, where he studied medicine and botany under Linnæus. He afterwards visited London and Edinburg, at which latter place he took his degree of Doctor in 1767. The thesis published by him on this occasion was "De Lavatione Frigida." Upon his return here he quickly rose to a high degree of estimation among his elder medical brethren. In 1768 he was chosen Professor of Materia Medica and Botany in the College of Philadelphia. In 1789 he was made Professor of the Theory and Practice in the University of Pennsylvania, and on the junction of these two schools was retained as professor of the same branches.

Of his writings we have nothing but his thesis, and a letter to Dr. Lettsom, published in the first volume of the Medical Society of London, on the diseases succeeding the transplantation of teeth. He took an active part in the establishment of the Philadelphia Dispensary, and of the College of Physicians, and was an uncommonly skilful and successful practitioner. He died in 1817, aged 75.

William Currie was born in Chester County, Pa., in the year 1754. His father was an Episcopal clergyman, who had emigrated from Scotland, and intended to devote his son to the same profession as himself, and with that view had him carefully instructed in the Latin and Greek languages, besides giving him some knowledge of the Hebrew.

Upon arriving at years of discretion, young Currie relinquished all idea of the study of theology, and made choice of the profession of medicine. He became an apprentice of Dr. Kearsley, the elder, attended the lectures delivered in the Academy, afterwards united with the University of Pennsylvania.

At an early period in the Revolutionary contest, Dr. Currie entered the American army as a surgeon, and was

29 (225)

attached in 1776 to the hospital on Long Island, and afterwards to that at Amboy. On the termination of the war he commenced the practice of medicine in the town of Chester,* from whence, in 1790 or '91, he removed to Philadelphia.

In the controversy which arose after the visitation of the yellow fever, in 1793, and subsequently, Dr. Currie entered warmly. He was a decided contagionist, and maintained at first the disease to be of foreign origin, but at an after date became convinced that it could be originated at home, though he always firmly held that its propagation was owing to contagion.

In 1792 he published "An Historical Account of the Climates and Diseases of the United States." This volume, of which a second edition, with additions, was published in 1811, consists of a collection of materials for an account of the medical topography, soil, climate, and diseases of each State, and a view of the modes of treatment. It is a work of merit, and embraces the result of more than thirty years' experience of a most successful practitioner. Another work of Dr. Currie, entitled "A General View of the Principal Theories of

* He advertises (May 2, 1781, Penn. Gaz. and Journ. of that date) that he comes to Philadelphia after eight years' practice.

Diseases which have Prevailed at Different Periods of Time,"
appeared in 1815; and in addition to these he put forth, at
various times, publications upon the yellow fever, and a
treatise upon the epidemic bilious fevers of America. In
this latter, which is his best work, he recognizes the difference
between bilious remittent and yellow fever, and his remarks
are distinguished by careful observation and sound sense.*

In 1818 he sank into a state of fatuity, and so continued
till his death in 1829. As a classical scholar, an erudite
physician, and a successful practitioner, Dr. Currie occupied
a high standing among his contemporaries. In the works
of the ancients, as well as in those of a later date, he was
deeply read, and kept pace with the improvements daily made
in the various branches of his profession.

As an author, he was never popular. His style was dry
and tedious, and he had the misfortune to be always in opposition to the reigning doctrines of the day in regard to the
unity of disease, the non-contagion of fever, etc.

A prominent trait in his character was his love for satire,
and in defence of his opinions he sometimes resorted to it,

* Currie, "Dissertation on the Autumnal Remitting Fever," Philada. 1789.
P. Stewart.

when facts were either opposed to him, or altogether wanting, and this frequently led him in conversation to place in a ludicrous light the foibles of his professional opponents; throughout life, however, he evinced a stern integrity which prevented him from doing injustice even to those who differed most from him in opinion. (Vol. 6, p. 205, of Register.)

To Philip Syng Physick is in great measure due the credit of having given to the surgical practice of the Hospital, and indeed to our city, the particular impress which has for so many years distinguished it, and in connection with the Institution which fostered him, it will not be inappropriate to trace an outline of his life and services.

Dr. Physick was born in Philadelphia in 1768, and received his early education at the "Friends' Academy," then under the care of Robert Proud. After graduating at the University of Pennsylvania, in 1785, he commenced the study of medicine with Dr. Kuhn. At the age of 20 he was taken by his father to London and placed under the care of John Hunter, where he distinguished himself by his industry and devotion to anatomical pursuits. In 1792, after spending a winter at Edinburg, he received his degree from the University of that place, and in the same year returned to Philadelphia, and entered upon his professional career. In 1794 he was made Surgeon to the Hospital, which position he held till 1816, when he resigned, and was succeeded by Dr. Dorsey. In February, 1801, he commenced a course of lectures there on Medical and Operative Surgery, in consequence of a request to that purpose drawn up and signed by a very large proportion of the students attending the University of Pennsylvania, and from the success of these lectures the professorship of Surgery was separated from that of Anatomy in the University, five years subsequently, and he was elected to fill it. This Chair he held till 1819, when he was transferred to that of Anatomy, which he continued to occupy till a few years previous to his decease. He died in 1837, aged 69.

Ardent in his professional course, he spared no sacrifice to devote himself to such as confided themselves to his care; and fixed in his principles, no desire of notoriety, nor wish to appear as an originator or operator, ever induced him to recommend the knife in cases of trivial deformities, or in those of doubtful utility. In his lectures he attempted no oratorical display; they were all carefully prepared and written out, and were practical, and founded principally upon his own experience. In all his treatment Dr. Physick was energetic and unvacillating, and when operations were necessary, he was remarkable for deliberate preparation for them, sympathy for the sufferings of the patient, and great devotion to their after treatment, and these qualities combined with his candour on their probable results, and the unquestioned truth of all statements made by him to his patients and brother practitioners, tended not a little to elevate his reputation. Though eminent as an operator, yet it is not in this way that Physick's claims to distinction can be upheld. It was by transplanting the principles of John Hunter to his native clime, by insisting on them in all his teachings, demonstrating them to his students, adopting them in his practice, and thus avoiding the performance of rash operations, or such as were not really necessary. It was with a character such as this that Physick rose to occupy that niche in the temple of fame in which his country-men have unanimously placed him, and the endearing title of "Father of American Surgery," justly and universally bestowed upon him, will ever mark the benefits he conferred upon our profession and our country.

To appreciate the services of Dr. Physick, and determine what he did for Surgery, we must measure him with his contemporaries, and compare the state of Surgery at the time he entered upon the practice of it, with the condition in which he placed it. Of capital operations, previous to the year 1750, but comparatively few were performed here.

Dr. Physick wrote but little, and that little was altogether upon subjects of practical utility.

To the Medical Repository of New York he communicated the particulars of a case of hydrophobia (1802), and a paper on un-united fracture, wherein he proposed the treatment by the seton (1804); also on gum-elastic catheter with bougie point, &c., in the same, vol. 7, p. 35, 1803. In Coxe's Medical Museum he published (vol. 1, p. 55, 1804) a case of varicose aneurism. A paper on the use of blisters in the treatment of mortification (vol. 1, p. 189, 1805), and one upon luxation forwards of the thigh (1805, vol. 1, p. 428). To the Eclectic Repertory he gave an account of a new mode of extracting poisons from the stomach (vol. 3, 1813), and a recommendation of the use of animal ligatures (vol. 6, 1816).

To the Philadelphia Journal of the Medical Sciences he contributed a paper on the use of the double canula and wire in the treatment of hemorrhoidal affections (vol. 1), and one on carbuncle (vol. 2, p. 172), and in the American Journal of the

Medical Sciences, one recommending an instrument for the excision of the tonsils, and on relaxation of the uvula ; he was the first to point out the fact of pulmonic irritation being produced by elongation of the uvula. . . . . *

* End of the unfinished manuscript.

www.ingramcontent.com/pod-product-compliance
Lightning Source LLC
Chambersburg PA
CBHW021526210326
41599CB00012B/1398